Physics Ideas
and Experiments

A collection of "Notes" from
Past *School Science Reviews*

Compiled by:
Randal Henly

Published by:

The Association for Science Education
College Lane
Hatfield
Hertfordshire. AL10 9AA

Design and Layout: D. Melchior, ASE.

Printed by:

Piggott Black Bear Ltd.
Cambridge
© ASE 2006

ISBN: 0 86357 407 6
ISBN-13: 978 0 86357 407 8

Physics Ideas and Experiments

Introduction

Andrew Bishop was editor of the School Science Review from 1966 to 1996. During those thirty years, a wealth of useful material for the teaching of physics (and indeed of chemistry and biology too) appeared in the pages of the SSR, and it was felt, by the Publications Committee of the ASE, that it would be a pity for so much good material to slip into oblivion. This publication — Physics Ideas & Experiments — is essentially a reprint of a selection of these physics notes from the SSRs of Andrew's editorship, and is left as a memorial to a wonderful teacher, editor and person.

Some material from that time is now dated (e.g., experiments with thermionic valves) and other items are (unfortunately) no longer acceptable on account of Health & Safety regulations. Those that have been selected for inclusion should be found useful/helpful for most syllabuses, styles of teaching and ways of organising the laboratory. The material has been classified into a number of sections, so that teachers with particular interests can easily find everything that is relevant to that interest.

The items in this volume have come from about 50 contributors working in various levels of education — primary schools, secondary and grammar schools and universities, and to all of these the ASE is indebted.

Health and Safety

The 'physics notes' are reproduced here as they originally appeared in School Science Review. They therefore conform to the standards of health and safety considered appropriate when they were first published, not necessarily those of today. The Editor and Publishers explicitly do not warrant that the procedures given would be considered appropriate nowadays. It is the responsibility of all those using these notes to make their own risk assessments before doing so.

Contents

1. Mechanics

2. Heat

Contents

2. Heat (continued)

3. Optics

4. Waves and Sound

Contents

4. Waves and Sound (continued)

5. Magnetism and Electricity

Contents

5. Magnetism and Electricity (continued)

6. Laboratory Organisation

7. Miscellaneous Items

Contents

7. Miscellaneous Items

Principle of flotation: poser and solution

W. Oxley, Technical High School for Boys, Salford 6

A man sits in a boat containing a large boulder floating on a pond: he drops the boulder overboard. Does this result in the level of the pond rising, falling, or staying the same?' Such a question can provoke much thought in a class and really test their understanding of the principle of flotation; it is also useful because through investigating the problem experimentally a spectacular result can be obtained.

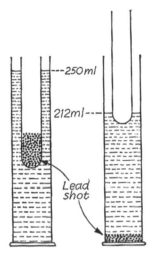

The situation is reconstructed by substituting in place of the boat a test-tube floating in a measuring cylinder, and by replacing the boulder by lead shot. The result is as shown in the figure.

A narrow measuring cylinder helps to ensure that the test-tube floats in an upright position.

A water barometer

E. G. Le Quesne, Thika High School, Kenya

This year I was put off the usual demonstration of the mercury barometer both by the propaganda about its danger to health and by the high cost of mercury. I found the water barometer a very satisfactory substitute, using a suitable branch of a tree in the school grounds as we have no buildings tall enough.

Twelve metres of clear PVC tubing (3 mm internal diam.) is marked with tape every metre. One end is tied to a weight and held under water (coloured) in a beaker. The tube is completely filled with water by sucking while the tube is still coiled. As water is siphoning out of the open end, a glass rod is pushed in to seal it. This makes a better seal than a screw clip. A string at least 24 m long is tied to this end of the tube and a small weight (2 N) tied to the end of the string. The apparatus is carried outside and the weight thrown over the branch by a pupil. He may need three or four tries! The tube can then be hauled up, and the final level of the water can be read easily from the ground if the water is quite strongly coloured.

On lowering the tube, a small length of air is seen at the end of the tube. This may be due to leakage or to the gases dissolved in the water (1.7 per cent by volume at 25 °C, our room temperature (!), is the value given in Kaye and Laby).

I find that this experiment stimulates a lot of questions and interest with junior forms.

A model demonstrating the movement of molecules in gases

A. V. Jones, Sunderland College of Education, Co. Durham

There are on the market many ingenious machines to demonstrate random movement of gas molecules and kinetic theory of gases. The increase of rate movement of molecules with temperature can be demonstrated with the simple model shown. The apparatus is cheap and needs no elaborate items of laboratory ware.

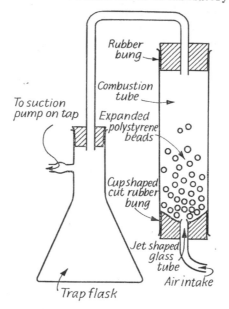

The expanded polystyrene beads were picked out of an old sheet of moulded polystyrene packing (a ceiling tile would also be suitable). When the suction pump is turned on, the beads begin to move and increased pressure enhances their rate of movement. Mixed coloured beads (made by dipping the beads in suitably coloured dyes) can show gas mixing and also enables the randomness of movement to be shown more clearly.

The number of beads and the size of the intake air jet are determined by trial and error, but the whole apparatus can be constructed and working in a very short while. Rubber bungs must be used.

Though not a perfect representation of what happens, this model nevertheless has much to commend it, not the least being its cheapness and simplicity.

'Streamline flow' analogy

W. H. Jarvis, Rannoch School, Perthshire

An interesting, though inexact, analogy with streamline flow can be produced as follows. Pour 200 to 300 pieces of lead shot, or suitably small phosphor-bronze ball-bearings, into a burette. Turn off the burette tap, and top up with glycerol. Seal by forcing a suitable rubber bung in, with the tap slightly open.

When the burette is inverted, the speeds of balls near the centre of the burette may be compared with the speeds of balls close to the sides.

A lawn-sprinkler to demonstrate jet-propulsion indoors

Alan Ward, St Mary's College, Cheltenham

A lawn-sprinkler jet-propelled by its own water supply is a well-known example of the reaction motor principle. Here are two ways in which this device can be operated as a teaching aid indoors, without mess.

Woolworth's sell a suitable model made in plastic, aluminium and brass, for less than ten shillings.

For the first demonstration, inflate a balloon, seal it with a clothes-peg 'clip', and fix the neck of the balloon on the sprinkler's inlet pipe. The result resembles a

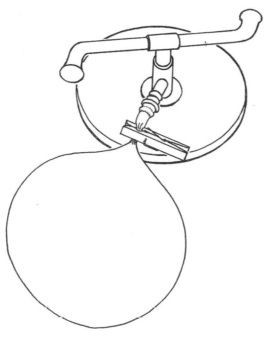

variety of bagpipes (see the illustration), but the 'motor' works splendidly when the peg is released and air forced from the balloon drives the 'jets' by reaction.

Alternatively, the apparatus can be connected to the main gas supply. Then, if gas pressure is high enough and friction in the machine low enough, the device can be gas-driven. Dramatize the effect by igniting the jets.

Principle of the turbine

C. J. Feetenby, Allerton High School, Leeds

A circular piece of metal, e.g. a tin lid, can be cut as shown in the diagram, into a number of sectors. Each sector is twisted until its plane is at an angle to the original

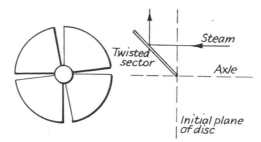

plane of the disc, making a fan. If this is mounted on an axle and a jet of steam from a small orifice is played on it the turbine principle can be illustrated. The change in direction of the steam is clearly seen.

An apparatus for the investigation of circular motion

A. K. Holmes, John Smeaton School, Leeds

Since circular motion receives considerable emphasis in present-day physics, it was considered that the development of a practical apparatus to show the relationship between centripetal force, mass, radius and angular velocity, would be useful.

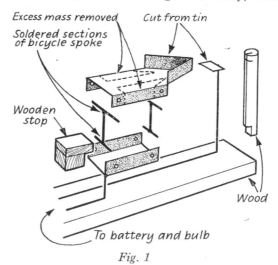

Fig. 1

A light hinged platform is mounted at one end of a lath (Fig. 1), at the other end of which is a post. An elastic band is attached to the platform, stretched to give the tension required, and secured to the post with a piece of copper wire. This holds the platform back against a stop. A known mass is placed on the platform and the assembly is mounted on a turntable (Fig. 2).

When the angular velocity exceeds a certain value, the platform will move outwards making a connection in a circuit which lights a bulb. This indicates that the angular

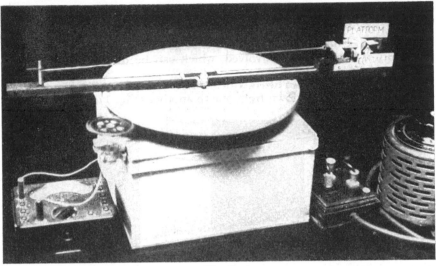

Fig. 2

velocity is such that the force in the elastic is barely sufficient to sustain the motion of the mass in a circle of this particular radius. Measurements can be made of the various angular velocities which are just large enough to light the bulb for different values of mass and tension, and the relationships between them tested.

Force: A paper strip stuck to the lath is calibrated in newtons using a newton spring balance to extend the elastic band which is fixed to the platform.

Mass: 10 g hanger weights are convenient if sellotaped together.

Radius: This is the fixed distance (20 cm) from the centre of the mass on the platform to the axis of the turntable.

Angular velocity: When the bulb is just lit, to time revolutions of the turntable is satisfactory, but a more convenient method is to drive a small d.c. motor acting as a dynamo from the turntable and to use the e.m.f. generated for calibrating a meter to read directly in radians per second.

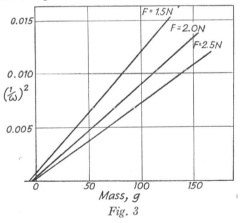

Fig. 3

An example of one of the sets of curves which can be obtained using this apparatus is shown in Fig. 3.

ACKNOWLEDGEMENT

The work for this note and the one following was done while I was a student in the Department of Education of Leeds University, and I should like to acknowledge the encouragement and advice of Mr W. F. Archenhold and Mr B. R. Chapman of that Department.

A simple approach to moments*

R. A. Hands, the Queen's School, Chester

It occurred to me after a recent derivation of Hooke's law by a class, using steel helical springs, that the same springs, by now familiar, could be useful in introducing the concept of a moment. This is the set-up that resulted, as an alternative to the traditional balancing of metre rules pivoted at the centre. The idea of turning something against the restraint of a spring, to illustrate the concept, is not new—I remember its being described some time ago in *S.S.R.*—but it was I believe as a demonstration, using something like a large clock spring. This needs no special apparatus except the nail and the hole in the metre rule, and so can easily be done as a class experiment.

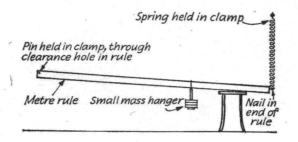

The instructions given are: Find various combinations of force (applied as the weight of some standard masses) and position of application which will just make the pivoted rule touch the tripod, despite the restraint of the spring. Then search for some quantity, derived from the force and position values, which is constant: this will be characteristic of the turning effect applied to the rule, and can be used to describe it.

The consistency of the moment values within a set has been to ±5–10 per cent according to the experimenters' competence. Further sets can of course be obtained by the use of different extensions of the spring, which will show proportionality to the moment values obtained. I doubt however whether this serves any really useful purpose.

Paper helicopters again—a puzzle

Alan Ward, Saint Mary's College, Cheltenham

Hold a completed helicopter (after folding out its wings), with the wings vertical (Figure 1). Now, when released, will the model (*a*) plummet and flutter down fast? (*b*) swoop down like a glider? (*c*) spin down fast? (*d*) open out wide (Figure 2) and spin

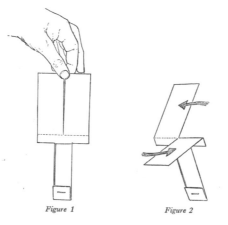

Figure 1　　　*Figure 2*

down normally and slowly? Remarkably, it opens out and drops normally. Why should this be so? As it starts to fall, air resistance forces it to spin. Then dynamic inertia of the upright rotating wings forces them to open. Dare we say that the wings are flung outwards by 'centrifugal force'?

Spinning stabilizes the falling helicopter, because spin makes the model behave like a gyroscope—keeping its axis vertical. The device provides an elegant demonstration in this lively area of physics.

Would it work on the moon?

Surface tension pulls a boat

Alan Ward, College of Saint Paul and Saint Mary, Cheltenham

Liquid detergent such as Fairy Liquid reduces surface tension on water. For a neat, easy-to-see demonstration, use a plastic plant-pot tray 'trough' (mine measures 15 × 60

cm) and a polystyrene food tray 'boat' (approximately 13 × 22 cm). Fill the trough with clean water and float the boat near one end. Wet a finger with detergent and just touch the surface of the water behind the boat. Detergent molecules quickly spread over the water, reducing surface tension behind the boat, and letting the boat be pulled along—almost the length of the trough—by the greater tension of the pure water molecules ahead. Wash everything before repeating the test. To make it easier to empty the trough, the demonstration can be mounted on supports that are tall enough to allow water to drain through a hole in the trough (kept corked during the tests) into a container beneath. By all means add some more fun to the display by putting a balsa-wood mast and a paper flag on the boat.

The cartesian diver

B. N. Arnold, Cornwall School, Dortmund

I am sure that most science teachers know the cartesian diver experiment using a winchester bottle and a semi-micro test tube or suitable alternatives. Pressure on the rubber bung compresses the air in the test tube allowing more water to enter the test tube thereby altering its 'relative mass' and causing it to sink. Releasing the pressure allows the diver to surface (see Figure 1).

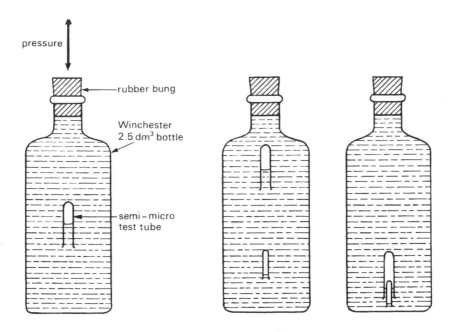

Figure 1. The cartesian diver   *Figure 2*

A worthwhile modification is to include a micro or ignition tube as well as the semi-micro test tube, the criterion being that one fits inside the other easily. The object of the exercise is for the diver to 'collect' the smaller test tube which travels at a different rate to the diver. This task is made all the more difficult by the water and the bottle refracting the light and altering the apparent position of one test tube relative to the other! The second problem is how to separate the two test tubes without removing them from the bottle. Both operations call for considerable manual dexterity and control (see Figure 2).

The experiment should give rise to several points of discussion, e.g. how the diver descends, why does the smaller tube ascend/descend at a different rate to the larger tube, why is it so difficult to separate the tubes once they are together, how does the bottle and water make it difficult to aim, how does tilting the bottle use gravity to help you position the diver (very awkward if you try to make the smaller test tube rise into the larger one!) etc.

A simple verification of the Principle of Moments

D. S. Newman, Balfour House School, St Athan, Glamorgan

Some pupils experience difficulty in verifying the principle of moments because it is not always easy to obtain a state of equilibrium with the lath in a horizontal position. The following arrangement allows for rapid and simple verification of the principle of moments in a way that eliminates this problem.

A half-metre rule, bored at the 1 cm mark, is pivoted on a strong plotting pin mounted in a cork. The cork and pin are mounted horizontally by clamping them in a retort clamp. A Newton dynamometer is suspended from a separate retort clamp and its height is adjusted until the other end of the half-metre rule that it is supporting is horizontal with the pivoted end. The reading of the dynamometer is noted, this has to be subtracted from later readings since it cancels out the moment caused by the weight of the rule.

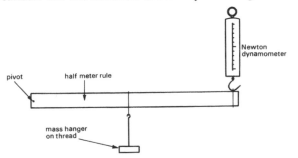

A mass hanger and mass, say total 200 g is then suspended by a cotton loop from the rule, the height of the dynamometer is again adjusted to bring the rule horizontal and the reading on the dynamometer is taken. The length from the pivot to the cotton loop and to the dynamometer hook are read off.

Finally the mass hanger and weight are suspended from the dynamometer to record the weight in newtons.

Clockwise and anticlockwise moments can then be calculated in the usual way.

One of the advantages of the method is that a large number of readings can be obtained using just the one mass hanger and weight at a variety of different distances from the pivot.

A model rocket to demonstrate Newton's third law of motion

Randle Hurley, Hextable School, Swanley, Kent

This extremely cheap model 'rocket' is both quick and easy to make. It can be used to demonstrate Newton's third law and, with a little ingenuity, can be used to provide data for momentum calculations.

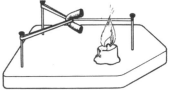

The rocket is cut from 1 cm thick wood. A light rubber band is attached to the two nails in the stern; the size should be chosen such that the band is under slight tension. The rubber band is tied back in the position shown by a loop of light string. A pellet of paper is placed on the band and the stub of candle is placed off centre as shown. The rocket is now ready to be placed in the water. As large a trough as possible should be filled to a depth of about 2 cm. The candle may be lit once the surface and the rocket are still and the rocket is pointing along the longest stretch of water. The candle is placed off centre so that the string does not burn through too quickly.

When the string burns through, the pellet is ejected and the rocket moves slowly forward. Resetting the rocket without a pellet and refiring shows that no movement results when there is nothing to throw away. The string should be wound round the bow pin a couple of times to prevent the string from being thrown off and thus upsetting the calculations.

The following experiments should get more mileage out of the rocket:

1. Weigh the pellet and the rocket, fire the pellet and measure the speed of the rocket. The speed of the pellet may be estimated using the data.
2. Repeat the experiment with a heavier pellet and then compare the speeds observed with the results of another pair of experiments in which the same two pellets are fired with a stronger elastic band. This data may be used as the focus for a discussion on the nature of momentum.

Demonstrating mechanical resonance

David Moore, Wellesbourne School, High Wycombe

The concept of mechanical resonance is required by many examination boards at both ordinary and advanced levels. However, textbooks, in the main, only refer to it in a brief manner. A typical approach might be to define the term and then refer to the inevitable child's swing and the Tacoma Narrows Bridge collapse.

I find the following demonstration both simple and effective.

The ripple tank motor with its already fitted eccentric wheel, is fastened to one end of a metre rule. A stout elastic band will suffice. The other end of the rule is firmly anchored to the bench, using a large mass. This arrangement, whilst not quite so rigid as obtained with a 'G' clamp, does allow quicker and easier variation of the length of the rule.

The natural frequency of the ruler can be established by 'twanging' it. The ruler is then subjected to forced vibrations from the motor. The frequency of these vibrations can be altered by using the variable voltage supply. Whilst the frequency is being altered the

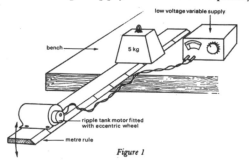

Figure 1

amplitude of vibration is observed. With low frequencies and high frequencies the amplitude will be small but, at one intermediate frequency (the natural frequency), the amplitude will be much larger as the ruler resonates.

The true significance of the often-seen (but not always well explained) amplitude/frequency graph can then be seen.

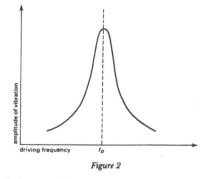

Figure 2

Some compromise is necessary, between satisfactory ranges of frequency either side of resonance and spectacularly large resonant amplitudes. Low frequencies give larger amplitudes but, at very low frequencies, the motor stalls. Experiment quickly determines the most satisfactory arrangement.

To establish that the resonant frequency and the natural frequency of the ruler are, indeed, the same, the apparatus can be viewed whilst illuminating it with a strobe lamp. The frequency of the lamp flashes is slowly increased until the vibration is 'frozen'. It will be seen that very little (if any) adjustment will be needed to the strobe frequency when moving between natural vibrations and resonance.

Helping to explain energy with a 'Boomerang Drum'

Alan Ward, College of Saint Paul and Saint Mary, Cheltenham

A popular item in my 'Energy Circus' of instructive demonstrations for primary school children, is the 'Boomerang Drum'. Many children are initially mystified by this amazing 'fun machine' which rolls away from, and then returns to its 'master', when he snaps his fingers. Sadly the whole purpose of the device is to reveal its magic afterwards, but the children enjoy it immensely, and are encouraged to build similar objects themselves. Here is the author's version of a very old idea:

The 'Boomerang Drum' is constructed from a 4 lb Cadbury's Drinking Chocolate container, measuring approximately 17 cm wide, and 15 cm diameter. It has a metal bottom, a cardboard side, and a conveniently transparent plastic top which fits tightly around the rim of the cylinder. When the top is on, it slightly increases the rim's diameter, so a second, identical top will later be fitted on to the rim at the bottom. (This will ensure that the device runs in a straight line when it is rolled along a floor.) Bore 0.5 cm diameter holes, 3 cm each side of the centre, in line along the diameters on both the plastic top and done to the spare plastic top. Use a nail to make the holes in the metal, and use a cork-borer to make holes in the plastic.

Two 10 cm long, thin rubber-bands are looped and fixed to the holed parts on each side of a 3 cm long Bulldog clipboard clip. When the Boomerang Drum is finished, the clip will have to be suspended on the total of four bands, to hang in the middle of the hollow cylinder. This is done by threading the free end of each band out through one of the four bored holes. Then, on the outside top and bottom, pairs of rubber-bands are linked together with a small metal paperclip. It only then remains to fix an old 200 gm mass in the Bulldog clip's jaws, before replacing the plastic top—while carefully aligning the holes between the ends of the drum. Finally, the diameters of the drum's ends are equalized, by fitting the spare top on the metal bottom.

I show the Boomerang Drum to the children (generally older juniors), as an example of an interesting machine which can store energy and then release it again. Having checked beforehand that the school hall's wooden floor is level, I roll the drum away from me,

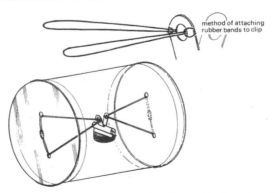

method of attaching
rubber bands to clip

Figure 1. How the rubber bands are fixed to the clip

across the children's line of sight. I am careful to let the children see only the opaque end of the device. After it has rolled for between 3 and 4 m, it stops. At that instant I snap my fingers. Then we all watch in silent fascination, as the Boomerang Drum begins to return, slowly at first, but then getting faster. After a second demonstration, I ask the children to guess how the machine works. I am accustomed to hearing a wide range of hypotheses, including 'magnets', 'plasticine stuck on the inside', 'water', 'weights', and 'a rubber band'.

We enjoy a brief discussion about what the cylinder might be like inside, before I take time to explain everything clearly. I can reveal how the parts are assembled by pulling off the plastic 'top' and then stretching the rubber bands. After replacing the plastic 'window', and by pointing its transparent surface towards the children, I can show them (by turning the drum in my hands) how the machine 'winds itself up' while it is rolling along the floor. The children can see the heavy mass hanging still (because of its weight), while the rubber bands are getting twisted. I explain how the twisting torque forces cause energy to be stored in the stretched rubber. This is the energy which drives the machine back to me, after the impetus produced by my forward push has been expended (by friction with the floor, air resistance, and by 'putting energy into the rubber motor').

Modelling energy concepts with a 'mousetrap bomb'

Alan Ward, College of Saint Paul and Saint Mary, Cheltenham

The gunpowder fuse—a trail of gunpowder leading over the ground—is like a chain of 'energy packets' waiting to be undone. This happens when a flame ignites the powder at the near end of the line. Burning gunpowder releases more heat energy, which ignites further particles in the gunpowder mixture. What we see is a sizzling lance of flame advancing towards a barrel of gunpowder at the other end of the trail. When the flame reaches its destination, the keg of dry powder catches fire with great suddenness. The reaction is instantaneous because gunpowder contains both fuel and oxidant. There is an almighty flash and bang, when chemical energy concentrated in the powder is released. That is how it happens in those old pirate movies on TV: an exciting example of a chemical chain reaction.

In my Energy Circus for upper junior school children, I illustrate the pirates' dangerous pyrotechnics with my 'mousetrap bomb'. During my discussion with the children, I let them see that a mousetrap is a nasty machine for concentrating and storing energy in a spring. The mouse triggers this energy when it steps on the platform to steal the cheese. My 'bomb' is simply a primed mousetrap with two polystyrene balls resting on the snapper. My 'fuse' is a line of dominoes, standing close to each other. The final domino (the 'detonator') has two others balanced sideways on top of it. The children appreciate that a 'packet' of energy is going to be released each time a domino falls against its neighbour. Now '5–4–3–2–1–Fire!' With a little prior judgement, I can usually get the flying balls to hit one of the children's teachers. Everyone enjoys the fun. . . .

Tiddlers and whales: some science with the fishing rod

David Smith, Huddersfield Polytechnic

There are millions of anglers in Britain, and almost any group of children will include devotees of the 'gentle art'. This interest can be made the basis of science lessons and investigative projects. Some possibilities in this connection are fairly obvious, for example the hydrobiology aspects, but there is also some potential for lessons on physics. The humble fishing rod embodies a range of physical principles, and these may be studied at a variety of levels, with plenty of opportunity for extension work for more able pupils.

One frequent question from children relates to how relatively weak lines may be used to catch relatively large fish. There are several points for investigation in this matter. First, fish usually *feel* rather larger than they really *are* (the 'one-that-got-away' syndrome). This may be explained by use of a simple test jig. Figure 1 shows a demonstration jig which consists of a 3 m bamboo pole (available cheaply from most garden centres). Small screw-shank hooks are fixed to the nodes. In the jig, the 'rod' is fixed to pivot at point F. There are force-meters at points E and L. The meter at E is zero-adjusted to compensate for the force due to the turning moment acting at point L. If a plasticine 'tiddler' is attached to L, then the effect of the turning moment produced by this load will be registered as a difference between the readings on the two meters. A small fish may be shown to feel big simply because it is in effect turning the angler's rod against him (or her). More able children may be set to investigate the effects of varying the angle at which the rod is set (effectively varying the perpendicular distance between load and fulcrum).

A subsidiary question relates to the importance of landing fish properly. It may be shown that the true dead weight of a fish will not be apparent until the fish is removed from the water. This is done by means of the test jig shown in Figure 1. The fish is immersed in a tank of water and then the water is taken away. Force-meter readings during and after immersion will show a considerable difference due to the upthrust which the fish received whilst in the water. This may serve to explain why so many fish are lost when they have been played out and brought to the bank. There is the possibility here that interested children may want to investigate the problem of fish buoyancy.

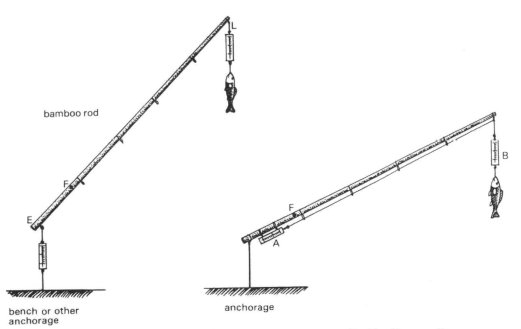

Figure 1. The 'one-that-got-away' test jig   *Figure 2.* Rod loading test jig

Additional Notes to Figures 1 and 2

It is difficult to give exact measurements for the various fixtures and fittings on the two test rigs: this is largely due to the variability of dimensions of bamboo poles. However, the following measurements relate to my own demonstration rig:

(a) The screw eyes are fixed at the nodes of the pole and vary between 25 cm and 30 cm apart
(b) The eye at point L is approximately 10 cm from the tip of the rod—the tip was not sawn off because the cane started to split
(c) The eye at point E is about 10 cm from the butt of the rod—these canes appear to be cut about mid-way between nodes, and the distances cited are not important
(d) Point F is 40 cm from point E (this was based on measurements taken from a Woolworth's fishing rod)
(e) The first screw eye is about 50 cm from point E

Possibilities now arise for rather more spectacular demonstrations. My colleague Mr Michael Shimmell tells of a demonstration in the school swimming pool where a 40–50 kg boy was landed using an ordinary general purpose fishing rod and 2 kg bs monofilament line. The volunteer 'fish' was instructed to make any manoeuvre which seemed appropriate, apart from directly snapping the line between his hands. The fish was landed with surprising ease!

The part played in all this by the fishing rod may be suggested by means of a modification to the test jig shown in Figure 1. A continuous line of appropriate strength is used. A force-meter (A) is placed at the point where the reel might normally be fixed (see Figure 2). A load is applied as before. The readings on force-meters A and B are recorded as the load is progressively increased. The reading at B represents the force on the line due to the struggles of the fish. Considerable discrepancies will be noted between the meter readings. The magnitude of these discrepancies will depend on the dimensions of the rod and on the exact nature of the cane.

So long as the load on the line does not exceed the line's breaking strain load, strain energy is being stored in the rod as it curves. In this respect the rod may be compared with a longbow. The fishing rod stores energy resulting from work done by either the fish or the angler, and this energy is then subsequently dissipated as work done against the fish. The unfortunate fish is truly hoist with its own petard. A further demonstration of this point may be made with the aid of a spring chest-expander: the child's effort merely leads to more strain!

Table 1. Some materials used in fishing rod construction

Ashwood	Whalebone
Crab-apple wood	High-tensile steel
Hazel	Aluminium
Yew	Glass-reinforced plastics
Greenheart	Carbon fibre-reinforced plastics
Bamboo { 'whole' cane / 'built' cane	

There are many important points which may arise from consideration of the construction of the fishing rod. The design of the rod is clearly the resultant of a number of compromises (for example length against lightness, flexibility against strength etc.). A variety of materials have been used in the construction of fishing rods, and some of them are listed in Table 1. It should be possible to acquire at least some of these materials and to compare their properties, perhaps with the aid of the test jig shown in Figure 2. A consideration of the properties and relative merits of rod-construction materials might lead to treatment of the special properties of composite materials, and hence to problems of civil engineering, biomechanics and so on: a long way from the water, and well into the realms of applied physics.

Storing leads

Peter Griffiths and John Stephens, Ladymead School, Taunton

Being dissatisfied with the normal peg board system for storing 4 mm leads we have developed the following idea. For lower school experiments it is unusual to need more than five leads. Therefore if the leads could be stored in sets of five they would be easy to issue and to check on return.

The storage method involved the use of a strip of plastic book binding material with a

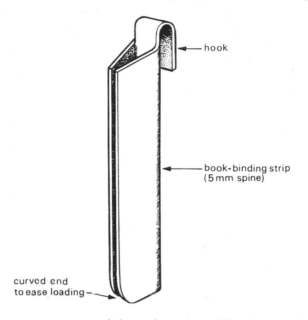

5 mm spine. This was cut at one end, heated gently and bent over to form a hook. The other end of the strip was rounded off to facilitate the loading of the leads.

To put a lead into the holder the lead was placed against the curved end and by running the thumb along the strip the lead is pushed in. This is repeated until all leads are in place, and the holder is then hung on a rack.

We have found it much easier to check leads for faulty plugs and have found that the method reduces plug loss and lead tangle almost completely.

The bicycle—a useful resource

J. W. Hawley, Inkersall Junior School, Derbyshire

Primary teachers must rely on everyday objects for resources to base their science lessons on since funds rarely allow primary schools to purchase expensive equipment. This is true particularly for apparatus concerned with studies about man-made artefacts and mechanical actions. The bicycle, however, is a valuable resource and one which is readily available. Most classes have a member who is prepared to bring his/her bicycle into the classroom. A model with derailleur gears is best suited for this purpose.

The following notes are some ideas for science lessons arising from consideration of a bicycle:

Structures
The frame is basically made up of two triangles which are strong shapes. Children can test this idea using meccano.

Forces
Consider pulling, pushing and twisting forces, e.g. pulling brake lever, pushing down on pedal, twisting the handle-bars.

Friction
Areas of low and high friction can be identified, e.g. wheels turn easily (low friction), brake blocks 'rub' on wheel rim (high friction).

Materials
Bicycles are made up of metal, rubber and possibly plastic components. Why do the properties of each material make it suitable for the components function e.g. strong metal frame, elastic rubber tyres, light plastic mud guards?

Movement
Consider the circular movement of the pedals, chain, rear sprocket and wheels.

Speed
The distance/time relationship can be studied following 'time trials' on the playground.

Gears
Count the teeth on each cog. What effect does this have on the distance travelled per revolution of the pedals?

Levers
Brake levers, gear levers, the steering mechanism and the pedals are all examples of levers.

Air
Consideration of the pneumatic tyre leads on to pumps, valves and air pressure.

Electricity
The cycle lamp provides us with an example of an electric circuit. The power source may be a battery or a dynamo.

Light
Whereas the lamp is a source of light the reflector uses light from another source.

Sound
The bicycle may have a hammer operated bell or an electric buzzer.

No doubt the reader can think of further possibilities for studies arising from consideration of the bicycle.

Indoor rocketry that juniors can understand

Alan Ward, College of Saint Paul and Saint Mary, Cheltenham

The most familiar examples of Newton's law that 'every action has an equal and opposite reaction' are the jet engine and the rocket motor, yet, personally, I have always found these more difficult to imagine than—for instance—the idea that it is the reaction of the firm ground to the downward action of my foot which supports and propels my walking. I suppose the reason for this difficulty is that, with jets and rockets, I am too distracted by the impressive surges of the exhaust gases, and therefore I find it difficult to concentrate on how the gases inside the devices are pushing 'forward' on them. Of course the example of the rocket is different from walking. The *action* of the gases pressing forward on the vehicle drives it, and the awesome effulgence that misdirects my attention is the *reaction*. I am well aware that what I have just said is the reversal of what textbooks generally say, but does it matter if action and reaction are equal as well as opposite? It also occurs to me that I may have stumbled on the reason why so many people believe that it is the gases pressing against the surrounding air that drives a rocket: a misconception, belied by space travel, which teachers are quick to dispel, without much corrective clarification.

These abstractions mean little to most primary school teachers, many of whom may be called upon to help juniors to understand how jet and rocket propulsion work. I would advise them to put aside what they may know about Newton's third law of motion—and to consider the simple explanation of why a toy balloon whizzes about, when inflated and then let go. Older juniors can appreciate that the air inside a blown up balloon is under pressure and being forced against the inside of the balloon envelope. (The energy in our 'balloon rocket' is stored in the 'spring' of compressed air, and in the tense 'stretch' of the balloon skin.) A diagram of a static balloon—with its neck shown sealed—can be used to show very

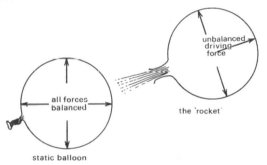

clearly how all forces acting on any part of the inside of the balloon must be balanced by equal forces acting the other way. A second diagram can be drawn, to show what happens when air is allowed to escape, after the neck is opened. Obviously the unbalanced internal force, acting on the inside 'front' of the balloon, must be the driving force. In my experience juniors can understand this idea.

When, during my Science Happening on 'Falling and Flying', I have discussed the balloon model with juniors, I am ready to stage a rocket launching as a finale. The twofold aim of my demonstration is to reinforce the explanation through some fun and excitement, whilst giving the attending class teachers a basic device for extending the science work

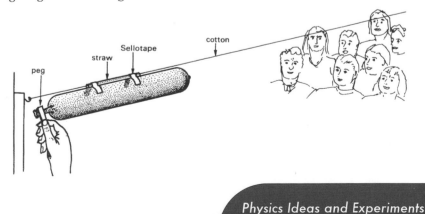

discard the needle, and tie the end of the cotton to my table. Then, when the line is tight, I have a child help me to fix the balloon underneath the straw, using two short strips of Sellotape. I make sure that the straw is over the top middle part of the balloon, and that the balloon and straw are in line. I undo the peg and grip the balloon neck—waiting . . . The children need little encouragement to begin counting down: 5, 4, 3, 2, 1, Fire! As the rocket balloon shoots up the line, the children react with happiness, wonder and a pleasing tumult of applause. And we can do encores . . .

Puzzles solved by inertia

Alan Ward, St Mary's College, Cheltenham

Dynamic inertia is neatly demonstrated by two ping-pong balls encapsulated between the cut-off tops of two transparent plastic squash bottles, which are simply tele-scoped together. The apparatus can be presented as a puzzle. Can the children find a way to put a ball at each end of the capsule? Of course, the answer is to spin the thing on its side, like a propeller—when inertia forces the balls to travel at tangents to the circular motion. This they cannot do inside their double conical enclosure, so

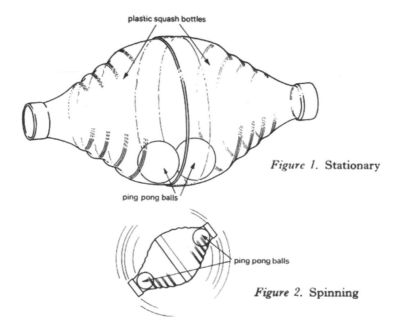

Figure 1. Stationary

Figure 2. Spinning

they roll to the ends of the device, where they remain until the spinning stops.

When we posed this problem on Birmingham's Dave Jamieson late-night radio show, an ingenious listener phoned in a unique solution based on the particular construction of the apparatus. He suggested that the bottle caps be removed from the capsule's ends. Next I was to point the apparatus upward, and to suck hard through the opening on the 'lower' ball. Then, while still sucking I was to invert the system—thereby letting the other ball drop, and therefore solving the puzzle!

Another inertia puzzle utilizes a goldfish bowl and a ping-pong ball. After placing two chairs about 2 m apart in front of the children, I tell them that the chairs represent Earth and Mars. On one chair ('planet Earth') I invert the bowl over the ball; while explaining that these objects signify an astronaut and his spacecraft. The puzzle is to get the astronaut to 'Mars' without turning the spacecraft upright, and by using nothing but the hands—which must not cover the hole in the 'bottom'. Readers will no doubt be able to appreciate how inertia can 'do the trick' . . .

Section 1: Mechanics

Putting tuppence on the 'Wall of Death'

Alan Ward, College of Saint Paul and Saint Mary, Cheltenham

Whirl around a glass marble inside a basin. The faster the marble travels, the more it spirals up around the inside wall—in opposition to the pull of gravity. Then, by giving the basin an extra strong jerk, the ball can be made to whizz out over the edge, straight across the room. The law of inertia says that moving objects must keep travelling along straight paths at constant speeds, unless checked or deflected by forces. Momentum (dynamic inertia) carries the bulk of the marble along.

Ignoring friction, the freely rolling marble exerts two forces: 1. A push arising from the basin's resistance to the marble's inertial motion forward ('centrifugal force'). 2. The push of the marble's weight, a consequence of the basin's opposition to gravity. The resultant R, 'compromise effect' of these forces makes the ball roll up the outwardly-sloping wall of the basin, until the resultant force can act at right angles to the basin's surface. See A and B in the diagram.

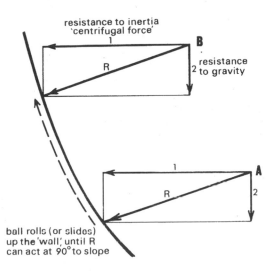

Forces exerted by the marble (ignoring friction)

The demonstration helps to explain how a brave and clever motor cyclist can drive his machine up around the increasing slope of the wooden, wall-like interior of the fairground's Wall of Death. Going at top speed around the vertical part of the wall, the rider is safe—providing that he does not steer above a horizontal path. To descend, he decelerates gradually, letting his decreasing resultant force be supported by the gentler slope lower down. (Of course all this would be impossible without motorbike tyre friction with the side of the wall.)

To make a twopenny coin ride the 'wall' inside a large pudding basin, hold the basin between both hands and shake the coin, until it wobbles up on to its rim. Then begin to revolve the basin, by moving both hands together in a circle. Soon the noisy clatter of the wobbling money should change to a regular rolling sound, as the coin hurtles like a motorbike wheel around the interior of the bowl. Acquiring the necessary skill will take a little patience. Try to get two coins riding the 'Wall of Death'.

Using your pupils as part of the experiment

Hester Greenstock, North London Collegiate School

1. TO FIND THE CENTRE OF GRAVITY OF A PUPIL

This is an experiment which at least the pupil involved will never forget. The equipment needed is a plank about 1.5 m long (a trolley runway does very well), a wooden block, and some bathroom scales, preferably calibrated in Newtons, on which the pupil can be weighed.

The first part involves weighing both the plank and the pupil using bathroom scales. The plank is supported at its ends by the block and scales respectively. By taking moments about the block end of the plank it becomes evident that: (i) the scales exert an upward force on the runway, and (ii) the centre of gravity of the plank lies at its centre (at least it does on ours).

The pupil then lies face up on the plank with her arms by her sides and soles of the feet above the wooden block. The scales show an increased reading, moments can be calculated about the block once again, and the pupil's centre of gravity found. Furthermore it is obvious that raising the arms above the head raises the reading on the scales and hence the centre of gravity must also have risen.

2. TO MEASURE ATMOSPHERIC PRESSURE

Our Magdeburg hemispheres happen to have a circular cross sectional area of about 50 square cm. It also happens that the average second or third year pupil weighs about 500 Newtons. (THEY think that they weigh 8 stone.) The effective pressure that they can exert across the area of cross section with their weight is therefore 10 N per sq. cm. or atmospheric pressure.

The Magdeburg Hemispheres are sealed and evacuated, and the usual attempts made to pull them apart. The upper hemisphere is then suspended from a beam so that the lower hemisphere has its handle just above the pupil's head. The 500 Newton pupil grips the handle of the lower hemisphere. They gradually support their weight from it and, as their whole weight is supported, the hemispheres part and the pupil does not have too far to fall. Ideally it should be shown that a 6 stone pupil CAN be supported.

Because of considerations about the directions in which the pressure of the air can act this could lead one into deep water with too sophisticated a class. However, the principle is sound, and it affords a dramatic and roughly quantitative measurement of atmospheric pressure to take one a stage further than the traditional collapsing can.

Introducing the topic of energy in the fifth grade (J4)

Alan Ward, College of Saint Paul and Saint Mary, Cheltenham

We began with 'hypnotized coconuts', a stunt I remembered from a Paul Daniels Magic Show—based on an ancient idea. Twenty puzzled fifth graders watched me tie twin coconut pendulums, a metre or so apart, on a green rope rigged across the classroom. The children could appreciate that the 'system' consisted of the rope, its window-catch supports and the pendulums, each about 120 cm long. What would we 'put into' the system if I raised one of the coconuts, then let it go? I sought for and received from the children the word 'energy'.

I defined energy as 'the go of things that makes anything work', but, 'not even the cleverest scientist really knows what energy is . . .'

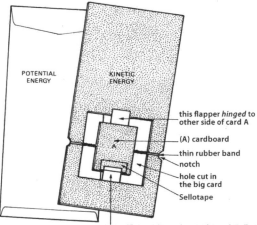

paper flapper *hinged* to card A with Sellotape

supports) and 'air resistance', the latter idea coming more slowly. A bright boy said 'aerodynamic flow', so we had a brief digression on streamlining cars.

Diverting attention away from the coconuts, I improvized another pendulum from scissors and string. A girl held this while I talked about doing some work on the scissors (using my muscles to lift the scissors against the pull of gravity) and thereby giving them gravitational *potential energy* (energy available for use). Letting go of the scissors would change potential energy into *kinetic energy* (from the Greek word 'kinetikos': to move). I gave the word 'cinema', formerly 'kinema', and the children realized the logic of calling film shows 'the movies'.

My main assignment that morning was to show a group of American fifth grade (J4) teachers some ways to enliven their teaching of the worrying energy topic, by making the ideas fun and by connecting the science with other subjects (language arts and craft work).

All I expected the children to remember were my definition of energy and the distinction between potential and kinetic energy. In passing, we discussed the potential 'go' of fuels, such as petrol that makes cars go. To reinforce these ideas I showed the children how to make a toy. From a long envelope marked 'potential energy' I pulled out a card on which was written 'kinetic energy'. As the card came out from the envelope it produced a startling buzzing noise, made by a cardboard flapper, driven by a wound-up rubber band, inside a hole cut in the card (stored energy changing into motion energy).

For half an hour the children were engrossed in making these toys, using the stiff cards, envelopes and rubber bands I had provided. I suggested that the children make joke Saint Valentine's Day cards, and I was pleased by their enthusiasm to draw hearts and flowers and to stick on funny gummed cutout figures which they took from their desks. As snack time approached I called for quiet, to ask my key questions. I was delighted to notice many more girls' hands up.

Then I showed the children a device (obtainable at stationery and toy shops) called a 'party popper', a small plastic bottle-like object, made in China. They knew that it contained a small explosive charge of potential energy. After a countdown—which children always enjoy—I fired the charge. There was a bright flash, a puff of smoke, and a shower of blue and yellow streamers (propelled by kinetic energy) brought our shared experience to a conclusion.

POSTSCRIPT

I have not mentioned my ambitious error in mentioning how energy is measured in joules, as this now seems irrelevant. The day after my demonstration two boys called at my borrowed office at the American Community School, Surrey—to ask me for extra coconuts, to help them continue experimenting with the pendulums.

The pitot-static tube

Geoffrey Auty, New College, Pontefract

The figure shows a simple model made at no cost. The polycarbonate 'pop' bottle should be of suitable shape, preferably having a gently tapered neck and a round base. The original stick-on base is removed. Most of the tubing consists of three old thermometer cases, the clear polythene type. To connect them together, two short pieces of flexible polythene tubing were obtained from laboratory stocks, although rubber tubing could also be used to make these curves.

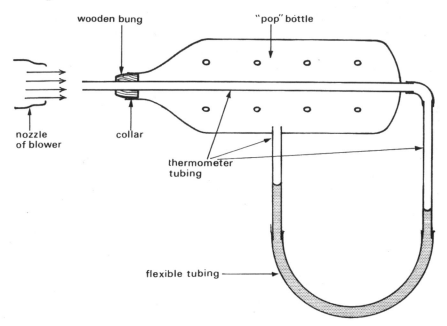

With care, it is possible to drill the large holes (approx 10 mm diameter) through which two pieces of thermometer-case tubing have to fit. Two rows of smaller holes (4 mm diameter) can then be drilled along the bottle. These are to equalize the static pressure. To seal the neck, I made a wooden bung in the woodwork department, but it is also possible to cut a rubber bung to a suitable shape, or to mould the shape using plasticine.

The system is held in clamps gripping the two vertical tubes. Water is added until these tubes are about half full. Then the bottle is fixed on top. The resulting model is uncalibrated but it illustrates the principle and shows clearly that there is a difference in wind speeds from a hair dryer and a cylinder 'vacuum' cleaner arranged to blow (as used to operate our linear air track).

The great water-jet scandal

[By coincidence, two science notes exposing this error were submitted to SSR at the same time. In fairness to all the authors, an edited version of both submissions follows.]

Keith Atkin, of Richmond College, Sheffield writes:

At the present time, great emphasis is placed on the value of practical work in physics education, particularly in GCSE physics courses. Students are encouraged to try things out for themselves and see what really happens. It is

therefore strange and disturbing to find a number of misconceptions in the shiny new textbooks specially brought out for GCSE.

I should like to describe one particular misconception which appears again and again. I refer to the demonstration that fluid pressure increases with depth. The authors invariably give the impression that they have themselves performed the experiment, whereas it is clear that either they have never done so, or have perhaps refused to believe the phenomenon in front of them.

Figure 1 is typical of diagrams to be found in a wide range of books covering this topic [1, 2, 3, 4].

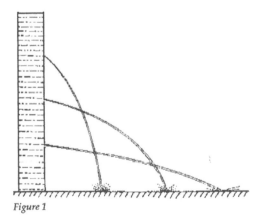

Figure 1

It shows a container full of some liquid such as water. Three holes drilled in the side of the container allow the liquid to emerge and follow curved trajectories. The liquid emerging from the bottom hole is depicted as having the maximum horizontal range. But this is WRONG. Figure 2 shows a photograph of the apparatus in action. The holes are drilled at one quarter, one half, and three quarters of the container's height. It is clear that the liquid emerging from the MIDDLE hole has the maximum horizontal range.

The argument traditionally presented in most texts is that pressure increases with depth and so the pressure at the bottom hole is greater than that at the other two. Consequently the liquid from the bottom hole shoots out the furthest.

The argument is, of course, fallacious. The flaw lies in the last step. While it is true that the pressure is greatest at the bottom hole with a concomitant maximum efflux speed, it does NOT follow that this gives a maximum range!

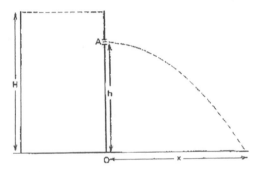

Figure 2

Derek Hart and Tony Croft of Crewe and Alsager College of Higher Education analyse the situation as follows:

A theorem in physics due to Torricelli[*] states that velocity (V) of a fluid emerging from an orifice in a cylindrical container, a depth d below the stationary surface is given by:

$$V^2 = 2gd$$

where g is the acceleration due to gravity. The direction of this velocity is perpendicular to the vessel walls. Consider the situation illustrated in Figure 3.

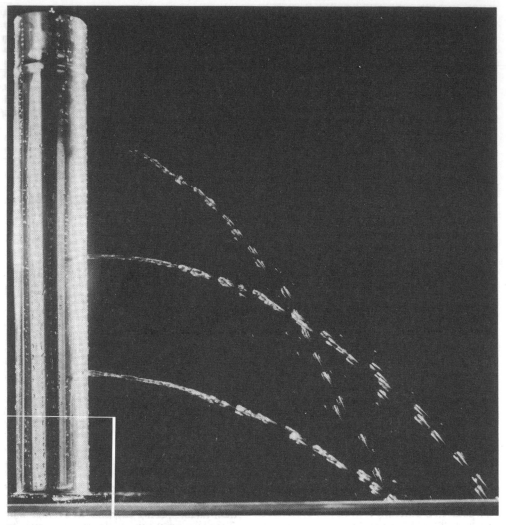

Figure 3

* Although Torricelli predated Bernoulli this result can be deduced from the well-known Bernoulli equation for fluid flow.

The constant depth of water is H. The orifice A is a height h above the horizontal floor.

Torricelli's theorem states that:

$$V^2 = 2g\,(H-h)$$

We now make an assumption that the stream of water can, to a first approximation, be treated like a projectile moving freely under gravity. On this assumption, if the jet strikes the floor at (where $0X = x$) then the time (t) for it to fall a distance h is the time to move a horizontal distance x.

(Note, that this explanation is true whether or not the vessel is raised above the horizontal floor.)

It is easily deduced from elementary mechanics that:

$$t^2 = \frac{2h}{g}$$

and therefore $x = Vt = 2\sqrt{h(H-h)}$

It follows that:

$$x^2 = 4h(H-h)$$

The *minimum* value of x^2 is zero (when $h = 0$ or $h = H$).

The maximum value is obtained from

$$\frac{d(x^2)}{dh} = 4H - 8h = 0$$

So x^2 (and consequently x) is maximum when $h = \lceil H/2 \rceil$

Section 1: Mechanics

That is the maximum range of jet occurs when the orifice is at the mid-point between the surface of the water and the horizontal floor [5]. If the container is placed on the horizontal floor then the maximum range does not occur from the lowest jet as depicted in Duncan [3] and Whiteley [4].

It may be surprising that a projectile projected with a high speed can have the same range as one projected with a lower speed. The bottom jet has the advantage of a high speed, but the top one the advantage of height, a crucial point. In Figure 2 the upper and lower jets are placed symmetrically so that this situation occurs. The 'velocity disadvantage' of the upper jet is compensated for by its height. A similar point in a different context has been raised by Hart [6] and Hart and Croft [7] where the advantage of a tall shot putter over a short one is discussed.

REFERENCES

1 Wellington, J, *Beginning Science: Physics* (Oxford University Press, 1984).
2 Avison, J, *The World of Physics*, (Nelson 1984).
3 Duncan, T, *Physics for Today and Tomorrow* (John Murray, 1978), p 92.
4 Whiteley, WL, *General Physics*, (University Tutorial Press, 1963) p 100.
5 Abbot, AF, *Ordinary Level Physics* (2nd edn) (Heinemann, 1969) p 121.
6 Hart, D, 'Standardizing the shot', *THETA* 1987, 1, 1, 3.
7 Hart, D and A Croft, 'Some thoughts on projectiles', *Teaching Mathematics and its Applications*, 1987, 2, 6, 71.

K Atkin, Richmond College, Spinkhill Drive, Sheffield S13 8FD.
A Croft and DA Hart, Crewe and Alsager College of Higher Education, Crewe, Cheshire CW1 1DU.

Steam–propelled 'airship'

Alan Ward

A picture of a jet-propelled steamboat (Figure 1) inspired me to get to work on a steam-propelled 'airship'. I aimed to make a model by minimal means. In the event I succeeded. On reflection I realized that the project involved many ideas relevant to science as it is taught in schools today.

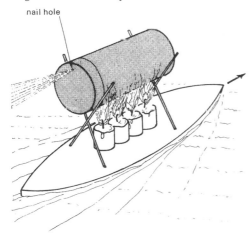

nail hole

Figure 1

THE IDEAS

Theory of levers and moments
Inertia and friction
Newton's third law of motion
Transfers of energy
Change of state - a little water making a lot of steam

Inefficient combustion - producing waste deposits of soot
Craft (skilful bending of wire)
Design (putting together a system for an intended result)
Technology (applying scientific principles economically to an artifact)
Problem-solving (getting a simple idea to work for you)
Safety (taking care not to harm persons or materials).

THE PROJECT (see Figure 2)

Make a double row by bending a 60 centimetre length of one millimetre thick iron wire (from a hardware shop). The bow must be longer on one side, because you need to twist it a few

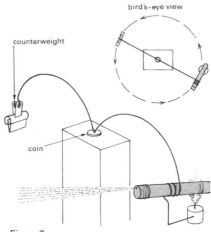

bird's-eye view

counterweight

coin

Figure 2

times around a lightweight metal film capsule, or a metal cigar tube - and fix a candle stub on the end of this side of the wire.

Cut the short piece of candle carefully - over a pad of newspapers - by sawing patiently with a sharp knife. (Warn children not to hurt themselves.) To ensure that the wire grips the capsule, prepare the wire by winding it around a thinner cylinder, such as a battery cell, or a fat pen. Use a match flame to heat the wire, where it will be attached to the candle.

Prick a hole with a darning needle, in the middle of the capsule's *metal* cap. This hole should be about a millimetre in diameter. Put a thimbleful of water inside. Let the capsule droop slightly towards its blocked up end. Balance everything on a coin that is resting on top

of a standing housebrick. You will need to find a suitable counterweight, to hook on the other end of the wire. Light the candle. Wait while the water boils. Act safely! *Always ensure that the hole in the capsule is not blocked.* (Presumably the slip-on lid of the tin can from which the original steamboat was made, acted as a safety-valve.) Then be rewarded by watching your model airship orbiting the brick.

At a recent public demonstration, the clothes peg I used as a counterweight held a paper cutout of Mickey Mouse. There were cries of disappointment from children in the audience, when after some minutes I stopped the show by blowing out the candle.

Alan Ward, College of Saint Paul and Saint Mary, Cheltenham

Go for a spin on a table top fun machine

Alan Ward

Obtain two identical press-on tin lids. Diameters should be no less than 9 or 10 centimetres. Put them together, rim to rim, to enclose a space packed with small marbles, all the same size. This device should make an effective ball bearing (Figure 1). Use it in a demonstration 'to reduce friction' - by spinning a heavy brick which you have placed upon it.

the model ball-bearing

Figure 1

More effectively, use the model ball bearing to spin a classroom table. Put the device on the floor in an open space. Get another person to help you to rest the centre of gravity of an upside-down table over the ball bearing. Let a small child sit on the inverted table. Readjust the balance of the arrangement. Then gently spin the table.

Recently I used this idea very successfully in a public demonstration (Figure 2). My young passenger in her new Brownie uniform enhanced the effect with the bright smile of enjoyment.

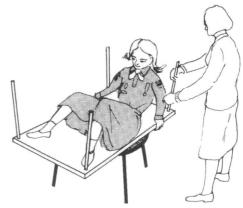

Figure 2

A Ward, formerly of the College of Saint Paul and Saint Mary, Cheltenham

Challenging an understanding of the amazing 'Wall of Death'

Alan Ward

Wondering what I could make from a blue plastic coffee jar lid, I hoped it might be possible to get it to roll around the inside of a big plastic bowl - in a demonstration of the principle of the fairground spectacle known as the 'Wall of Death'. For less than a pound, I had just purchased a large, red coloured bowl that tapered from top to bottom, like a pudding basin. I knew I could get a coin to roll around the bowl's interior, although some deftness and skill were needed.

I put the jar lid inside, before shaking the bowl to make the blue lid stand up on end. Then, by holding the bowl between both hands, I shook it with a quick circular motion.

Much to my satisfaction the coffee lid started to roll like a speeding motorcyclist around the inside wall... I found it much easier to get the coffee lid rolling than my previous experience with getting a coin to roll.

Inertia of the motion I gave to the roller made it tend to press forward along tangential paths to the circle of my revolutions. Of course it could not penetrate the wall of the bowl, so the interior surface acted on the roller with a centripetal force. This had the effect of preventing gravity from pulling the jar lid down. There was sufficient friction between roller and bowl, to make the jar lid rotate rather than slide along... With thoughts such as these, I could begin to understand a mildly challenging, but utterly fascinating stunt with practical physics.

A Ward, formerly of the College of Saint Paul and Saint Mary, Cheltenham

A cheap and simple gravity motor made in five minutes

Alan Ward

Use pliers to bend up 0.5 centimetres - through 90° at one end of a roughly 3 cm long metal paperclip. Use this clip to attach a thin strong thread to the centre of the top end of an empty ring-pull lemonade can. Take advantage of the hole in the can when you attach the clip to the metal top surface. Next make a circle of 8 holes, equidistant from each other, near the bottom end of the can. There might be a printed band of 'trimming' to help you get all the holes in the same level plane.

Bore the holes by using a fat darning needle. After making each hole, and with the needle point still inserted, push the needle over to the right - to lie, as it were, at the tangent to the circle of holes you are making. Therefore you incline all the holes to the right.

After filling the can with water, dangle it from the thread over a wide bowl. As the water drains away - in a dynamic pattern similar to a spinning firework - the can is forced by reaction to the jetting water, to rotate to the left. This example of Newton's third law of motion is a consequence of the rightward inclined holes. 'Uphill' positional potential energy of the water is transferred as kinetic energy, to drive a jet motor.

This gravity powered motor can be adapted to work a simple machine or toy.

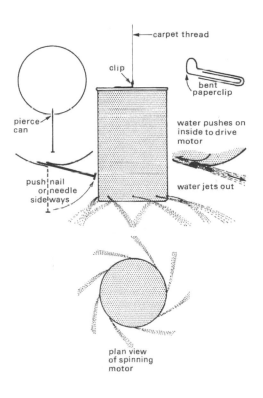

A Ward, formerly of the College of Saint Paul and Saint Mary, Cheltenham

The crumple zone challenge

Michael Brimicombe

Industry and government are always exhorting us to make A-level science courses relevant to the needs of the nation. The following practical encourages students to apply their knowledge and understanding of force, momentum and energy to the design of crumple zones in cars. A well-designed car incorporates regions which are designed to collapse in the event of an accident. The ability of a car to crumple in a collision is an important safety feature - the more it crumples, the longer the time during

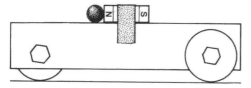

Figure 1

which the momentum of its occupant has to change and the smaller the restraining force has to be.

Figure 1 shows how the car can be modelled with a dynamics trolley. A bar magnet is taped to a dynamics trolley. The magnet holds a ball-bearing on the top of the trolley. The ball models a baby, the magnet models its safety harness and the trolley models a car. If the car slows down too rapidly, the harness will be unable to restrain the baby. Check this by allowing the trolley to run down a ramp into a solid object (Figure 2). Your task is to build a crumple zone from paper and card on the front of the trolley so that the baby survives a collision at an initial speed of 3 m s^{-1}.

The length l of the crumple zone is critical. You can estimate its minimum value as follows:

1 Measure the force needed to remove a ball-bearing from the end of the magnet - call this F. Measure the mass m of the ball.
   Use $F = \Delta p / \Delta t$ to find the minimum time Δt, needed to slow the trolley down from 3 ms^{-1} to 0 ms^{-1}.
2 Assume the trolley is uniformly decelerated during the collision. Use Δt to calculate the smallest safe stopping distance l for the trolley.

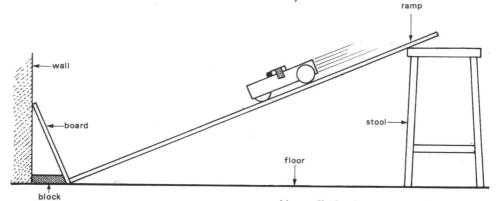

Figure 2

You will also have to consider the average force that the crumple zone will have to exert on the trolley to bring it to a halt. Finally, when you have built your crumple zone, you will need to calculate how far down the ramp the trolley must roll before it reaches 3 m s^{-1}. The easiest way of doing this is via a consideration of energy conservation.

A student who calculates the minimum length of crumple zone, includes a 50% safety margin and designs a crumple zone which can collapse a long way, and yet provide a large enough force will have the satisfaction of saving the baby when the crumple zone is tested.

M Brimicombe, Upper Cedars School, Leighton Buzzard, Beds

Simple technology with submersibles

Alan Ward

During the Second World War, when I was a child, I laid 'mine fields' in the River Exe and Exeter Canal. My innocent mines were balls of silver paper on cotton strings, tied to small stones (Figure 1). The stones sank, pulling down the balls (which contained air) to a depth of a metre or so beneath the surface of the water.

With that childhood memory, I invented the 'one-shot sub' (Figure 2). It sank to the bottom of a deep aquarium and then came up again, after leaving some of its mass behind on the bottom.

My one-shot submarine was a hollow plastic ball, perforated by a pattern of many large holes - a type of ball, common and cheap in toy shops.

This ball subsides slowly and barely floats. If a wire or a large paper clip - bent to form an angle of 40° - is hooked on the ball, the device keeps on sinking. But when the wire touches the bottom, it detaches itself and allows the ball to return to the surface.

Can this 'submarine' be modified to make it sink to the bottom, then rise - to stop in a position, say 30 centimetres underwater? Different coloured balls can be used to hold a submarine race - the first one down and then up again being the winner.

Recently I rescued a clean flask-shaped bottle from amongst the litter in a hedge. I used the bottle (Figure 3) to make a different sort of submarine... Get a thin, flexible piece of plastic tubing. Poke one end of the tubing deep inside the bottle. Sink the bottle in an aquarium tank, then get it to rise and sink, by controlling the amount of air you put inside it.

Alan Ward, 4 Branch Hill Rise, Charlton Kings, Cheltenham, Glos GL53 9HW

rubber band stops tubing coming out of the bottle

40°

Figures 1-2-3

Hooke's Law - again!

Mick Nott and Jeff Sturgess

Whilst we can't find Hooke's law anywhere in the new National Curriculum we are sure it must fit in to most schemes of work for children in the early stages of Key Stage 3. Recently we have both had experience of working with two Y7 classes on this topic.

The scheme of work in place did the usual 'stuff' of incrementally loading the spring, measuring its length with a ruler, recording the results in a table, plotting a line graph, drawing a

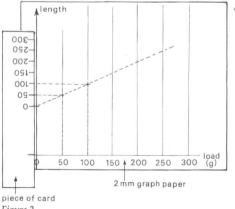

Figure 1

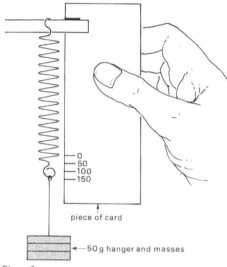

piece of card

Figure 2

line of best fit and then using this calibration to weigh things or predict the length of the springs for known weights.

The mixed ability groups we taught had some children who found the measuring to the accu-

racy of 1 mm tricky, others found the ordering of the results in a table required a great deal of concentration and then some had all the familiar problems of drawing scaled axes and translating the readings from the table to correctly plotted coordinates on the graph. And then we had to explain the line of best fit....!?

So one of us (JG) decided that the experiment should be turned around. The children were familiar with 'scales' or spring balances so the first practical experience became to make a scale. A piece of card (from an old box) was held next to the spring as it was incrementally loaded. The length of the spring was recorded on the card for loads from 0 g to 100 g in 10 g steps (see Figure 1).

Children were encouraged to try and identify a pattern, which came easily to most - the steps look roughly the same. Next we asked the children to weigh things using their scales. They could weigh things within the 100 g limit to an accuracy of 10 g and some of them to an accuracy of 5 g.

Then some tried to weigh things which weighed more than 100 g. Initially this was perplexing but then groups started to come up to us and say its 150 or 200! When asked how they knew we were told 'That's easy, we just count on the number of steps'. Brilliant! they were extrapolating; they had recognized the pattern and were making predictions from it.

The pieces of card were labelled with names and stored away. The other of us (MN) decided to use the marked card to directly draw the graph. After having their memories jogged about the work on coordinates they had done in maths, graph paper was issued and the children were given instructions how to draw the load axis in the x direction. They then drew the y-axis and placed their piece of card against it as in Figure 2.

The children were shown, using an example drawn on the blackboard, how to plot the y-coordinate from the card to the corresponding x-coordinate from the load axis. Two or three examples were shown and the result was 27 sets of plotted coordinates which all seemed to have pretty good lines of fit. Astonishing!; the most painless time that MN can recollect teaching it. The process of using the graph to weigh things to the accuracy of 1 or 2 g then became 'naturally' the reverse of the plotting exercise above - wonderful.

SUMMARY

We think the sense of achievement of making a 'working' scale was a very positive one for the children. We also think that by passing the technical and manipulative skills of proficiently using a 30 cm or 50 cm rule and the necessity to record anything in a table brought the graph closer to the experience with the spring. The graph was related to the spring not to the table as in most schemes of work. If your educational objectives are the establishment, recognition and use of a pattern then we recommend that you give the above method a try when introducing youngsters to Hooke's Law.

M Nott and J Sturgess, Centre for Science Education, Sheffield Hallam University

What a squirt!

PJ Sapwell

A simple variation of the age-old experiment illustrating that pressure in a liquid is exerted in all directions equally.

The cost of the apparatus is negligible, compared to £20 for a Pascal syringe made in glass.

Doing the experiment in a large bowl enables everyone to see the jets without creating the chaos of water everywhere!

The light (of the OHP) underneath illuminates the bowl and greatly enhances the effect.

PJ Sapwell, Beeston Hall School, West Runton, Norfolk

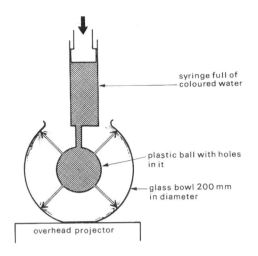

syringe full of coloured water

plastic ball with holes in it

glass bowl 200 mm in diameter

overhead projector

Amazing boats - ideas for the experimenter

Alan Ward

When I was a boy I saw a picture on the cover of a comic, illustrating the impossibility of being able to lift yourself by pulling your boot-laces... A sailing vessel that goes by blowing its own wind would seem to be a similarly impossible proposition. Some years ago I came across a drawing of a toy boat that did just that (Figure 1).

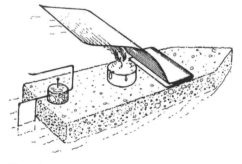

Figure 1

Hot air rising from a candle stub is deflected backwards by a 'sail' made of thin bent metal. Where the air is deflected it must push forward on the sail - and so it drives the boat. To test the idea I made a prototype model from a hollowed chunk of polystyrene, measuring $3 \times 8 \times 25$ centimetres - part of the packaging for a garden tool (Figure 2).

Although I have been familiar with the principle for many years, this was the first time I had

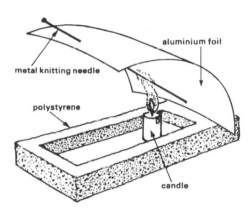

aluminium foil

metal knitting needle

polystyrene

candle

Figure 2

tried it out... My candle-powered sailing boat worked - albeit its voyages in my bath were slow. I was also remembering some similar boats described in *The Boys Book of Mechanics and Experiment*.

Figure 3

SAILING WHEEL

Let this wonderful design (Figure 3), adapted from a picture in Charles Ray's book, be an inspiration for your students: it could be floated safely in a paddling pool.

Make the disc from a large ceiling tile. Shape the fireproof sails from aluminium foil. When folding the foil, before cutting a sail with scissors, it is a useful tip to slip a piece of scrap paper between the folded parts (then the cut edges of the foil will not stick together).

Shape the candle stubs carefully - over a pad of newspapers - by patient sawing with a sharp knife. *Don't forget the wicks*.

The candle-powered floating wheel would be an unusual and attractive item to demonstrate at a Science Fair or Open Evening.

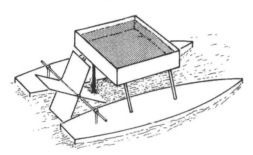

Figure 4

POSTSCRIPT

It goes without saying that experiments involving flames are dangerous and should be supervised by responsible adults. Also, it should be remembered that heating plastics produces toxic gases.

Figure 4 shows another suggestion from Mr Ray's book. Water falling through a hole in the raised water-tank is supposed to drive a paddle wheel and propel this weird catamaran. Mr Ray also suggested that such a boat might be driven by a stream of falling sand... Amazing! *Would these ideas work?*

REFERENCE

1 Ray, Charles, *The Boys Book of Mechanics and Experiment*, (The Amalgamated Press Ltd, *circa* 1937).

A Ward, 4 Branch Hill Rise, Charlton Kings, Cheltenham, Glos GL53 9HW.

Measuring pressure

N Vandyk

To the best of my knowledge, there are few - if any class practicals to demonstrate the effects of pressure, and I have found this one to be very useful.

An ideal pressure measurer would consist of a large number of identical springs, each fixed at the base to a flat surface. When a force is applied to an area covering the top of one spring, it is compressed by an amount proportional to the force. If the same force is applied over an area covering two springs, the force on each spring is halved, and both springs will be compressed by only half the amount (Hookes Law). Thus the distance each spring is compressed is proportional to the pressure.

An alternative to using a set of identical springs is to use a cheap polyurethane sponge. I recently bought a pack of 5 for 79p at a market stall. Alternatively, a large 'car washing' sponge can be cut in half.

To turn the sponge into a set of springs, use scissors to cut the sponge to about three quarters of its depth, lengthwise and across into approximately 1.5 cm squares (see diagram).

The practical was designed for a Year 9 class. Each group was given a 'modified' sponge, a 10 N (1 kg) weight, five plastic beakers ranging from 4.5 to 10 cm diameter and a 30 cm ruler.

The task was to find out how well the sponge performed as a pressure measurer by plotting a graph of the distance the sponge was depressed versus the applied pressure.

The beaker was placed on the sponge right

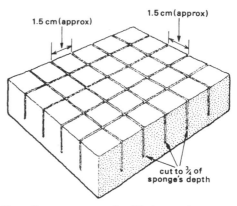

1.5 cm (approx)   1.5 cm (approx)

cut to ¾ of sponge's depth

Figure Pressure measurer 'modified sponge'

way up, and the distance between the bottom of the sponge and the beaker's rim measured with the ruler. The measurement was repeated with the 1 kg mass placed on or in the beaker, depending on the beaker's diameter.

The practical was attempted with a top and a bottom Year 9 set. The top set were asked to calculate the area of the base of each beaker by measuring its diameter. With the bottom set, the area of the bases were given.

The graphs obtained were reasonably linear and the majority of both classes found the practical an enjoyable and worthwhile exercise.

The practical could be extended by using different masses to show that doubling the area produces the same effect on the sponge as halving the mass.

N Vandyk, Ashmole School, Southgate, London N14

DIY piston and cylinder

Geoff Auty

Two articles in the December 1994 *SSR*, [1, 2] prompted me to suggest a simple way to obtain a piston and cylinder of good quality. Many DIY products such as fillers and sealants are now sold in plastic tubes designed for use with an injection 'gun'. The output end has a screw-on nozzle whilst in the open end there is a plastic piston which is extremely well fitting, intended to be pushed along by the action of the 'gun' to extrude the contents from the nozzle.

Having just emptied one of these tubes, and before the contents had solidified, I pushed out the piston by inserting a long steel rod through the nozzle. I have to admit that considerable force was needed to start it, presumably because some filler had dried on the inside wall of the tube, and resorted to a gentle tap of the rod with a hammer. Then I cleaned out the remaining filler, taking care not to damage the inner walls. A steel ruler proved to be a very effective cleaning tool.

To make this into a usable pump, the piston needed a push rod. I used a piece of wooden dowel rod 22 cm diameter. After drilling the plastic piston in three places, it was fixed to the end of the rod using self-tapping screws. As the piston was such a good fit, having ridges to produce the effect of piston rings. I used a fine file to trim it gently in order to make the movement in the cylinder easier. The assembly was then as shown in the Figure 1.

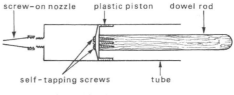

screw-on nozzle   plastic piston   dowel rod

self-tapping screws   tube

Figure 1   Assembly of pump

An alternative would be to use a steel rod about 5 mm in diameter and cut threads to take nuts at each end. A plywood disc could then make a suitable handle. A further refinement would be to add another plywood disc at the tail end of the cylinder to act as a guide for the rod. This needs to fit the plastic tube reasonably well to avoid distortion. If available, a quicker and better solution is to use the piston from another tube of the same size. The result appears rather like a bicycle pump but with a tight piston rather than the cup washer which acts as a valve.

The finished pump can be used like a syringe, or valves could be fitted as described by Alan Ward [1].

The popular size for DIY products is a tube of 46 mm internal diameter and 210 mm length. To demonstrate the principle of the hydraulic press or hydraulic brakes, two or more of these arrangements are needed, and preferably with different diameters. There are, of course, a number of cheap small syringes available from education suppliers, but various other products such as toothpaste are now available in similar

tubes of various sizes, and I found one example of a sealant in tubes of 145 mm length and 28 mm bore.

 SAFETY NOTE

Materials sold for the DIY market are not generally dangerous but read manufacturers' instructions regarding the contents of these tubes before attempting to dismantle. Fillers are generally slightly soluble in water and can be washed from hands. With sealants, other solvents will have to be used so rubber gloves and appropriate ventilation are advisable. Before starting, consider whether the use of such solvents is allowed.

REFERENCES

1 Ward, A, 'Cardboard engineering - making a force pump', *SSR*, 1994, **76**(275), 89.
2 Sapwell, PJ, 'What a squirt!', *SSR*, 1994, **76**(275), 96.

Geoff Auty, New College, Pontefract, West Yorkshire

Black objects absorb radiant heat more readily than shiny bodies

R. L. Page, St Paul's College of Education, Cheltenham

For this demonstration two identical bottles or flasks are fitted with corks. One of the bottles is blackened over a candle and then both are filled to an equal depth with coloured water. The levels should be about half way up the bottles. Each cork is bored and fitted with a narrow length of glass tubing about 20 in long. The corks are then fitted into the necks of the bottles so that the glass tubing reaches below the surface of the coloured water and pressed home until water appears in the glass tubing above the corks. The corks are adjusted till the levels of the water in each tube are the same. The bottles are then placed in the sunlight to get equal quantities of sunlight and the levels of the water noted after an hour or so. Alternatively the bottle can be placed in front of the filament of a bar electric fire about 9 in away from the filament.

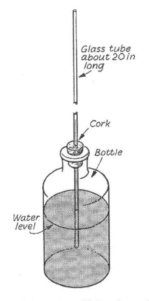

Glass tube about 20 in long

Cork

Bottle

Water level

The water level in the blackened bottle will be found to be higher than in the other bottle, showing that the air has expanded in the blackened one more because it has absorbed more radiant heat.

An improved method for the demonstration of convection currents

Ch. Ramakrishnayya, Raja Rao Bhavayamma Rao Government Higher Secondary School, Pithapuram, India

The conventional methods for demonstrating convection currents in water, using a crystal of potassium permanganate, have considerable limitations and, after trying out a large number of variations, I have found the following to be consistently successful.

A small crystal of potassium permanganate was enclosed in a small gelatine capsule purchased from a medical store. To keep it under water, it was tied to a small cube of lead. The arrangement was gently put at the bottom of a beaker half full of water, and the flame of a spirit lamp was allowed to play on the part of the beaker immediately beneath the capsule. After some air had bubbled out, a pink streak rose clearly upwards and the usual phenomena were very easily observed. Alternative methods for sinking the capsule were also used with success, e.g. inclusion of two or three lead shot in the capsule.

Cheap calorimeter cases from waste polystyrene

P. T. Haydock, Andrew Cairns Secondary School, Littlehampton

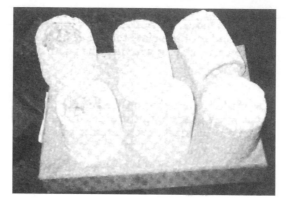

We used polystyrene sheets that came as packing with apparatus such as balances. Rings were cut from the polystyrene with a hot wire. The rings were then built up to the required size and glued together.

The cases are found to be light, easy to use, and make very good insulators.

Lids to fit can be made in a similar way with holes for thermometer, stirrer and heating coil if required.

The construction of cheap solar panels

C. D. Tuck, Cornwall School, Dortmund

The heat exchanger panels from the back of discarded domestic refrigerators make excellent solar panels when mounted in a simple wooden box that has been insulated with expanded polystyrene ceiling tiles and then covered with glass.

The circulation of water is achieved by running a windscreen washer pump mounted on the reservoir at about 9 V. This reduced voltage lessens the chance of overheating while running for longer periods than would normally be the case as a windscreen washer.

The photograph shows two panels. This arrangement allows experiments with controls to be set up. Various experiments concerned with improving the efficiency of the panels were tried with a sixth-form CEE group, and some worthwhile results were obtained, in spite of the weather invariably either raining or clouding over whenever experiments were attempted.

Flying a hot-air balloon in a school hall

Alan Ward, Saint Mary's College, Cheltenham

The gas-fuelled apparatus shown in the photographs is designed to demonstrate the 'flight' of a paper hot-air balloon indoors, when bad weather prevents outdoor flying. This is a convenient means for guaranteeing immediate test flights whenever hot-air balloons are built in school. The launching can take place in a large hall where there is no gas supply.

A boss and clamp are used to fix a bunsen near the top of a high retort stand. The bunsen is connected, via a long rubber tube and a short metal pipe, to a big plastic beach-ball that is filled—in advance—with gas directly from a domestic main supply.

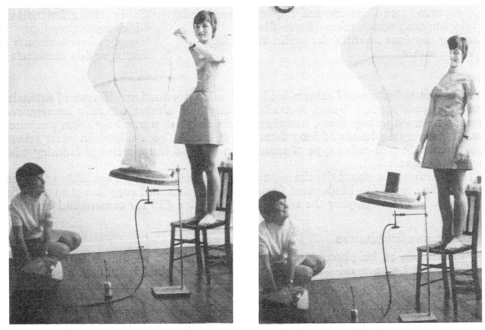

Fig. 1   *Fig. 2*

There is a 'control valve' in the form of a tight clip attached to the rubber tubing. Above the bunsen, secured by a second boss, is the combined chimney-lid of a cheap garden incinerator.

Steady pressure is applied to the gas bag, before the control valve is undone and the bunsen lit. Pressure upon the bag must be maintained by an assistant during the launching, and care must be taken not to let the flame come above the lower end of the chimney. Use the apparatus cautiously. When a draught of hot air is rising above the bunsen, the balloon is held with its neck around the chimney, until the paper is inflated and seems to be tugging upwards. Then, when released, the balloon ascends vertically and bumps splendidly against the ceiling.

The balloon shown in the photographs, which were taken by David Hall, is one of several made in an afternoon by thirty-six 10–11-year-old boys and girls in a class-room at Charlton Kings Junior School, near Cheltenham.

Specific latent heat of steam

W. H. Jarvis, Rannoch School, Perthshire

The simple method described consistently gives answers within $\frac{1}{2}$ per cent of the accepted figure.

The immersion heater is the 12 V Nuffield pattern; for this experiment it is run at 20 V, to give nearly three times the rate of heating.

We usually pre-boil the water to be used, as the immersion heater would take quite a long time. (Our tap water is tastefully impure.)

Readings do not start until boiling is proceeding steadily. Then the stopclock is started ($t = 0$) and the mass reading on the lever balance, m_1, is noted. It takes about 20 minutes for 30 g of water to boil away; the time and new mass (m_2) are noted after such an interval.

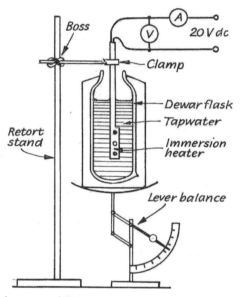

The formula $VIt = (m_1 - m_2)L$ is used, without any attempt at a correction for heat losses. There is of course an error due to the change in upthrust as less of the heater becomes immersed, but it turns out in practice to be negligible, as are the heat losses.

A cooling curve for water

John Marshall, Christ College, Brecon

A class experiment on the cooling curve of naphthalene can be improved by letting everyone plot two cooling curves, the second curve being for water. A mixture of 1.0 kg $CaCl_2$. $6H_2O$ and 0.7 kg pulverized ice gives a freezing mixture at about $-55°C$ [1], sufficient for about ten experiments at no very great cost. A small test-tube, one-third full of water at room temperature, gives a cooling curve from 15°C to $-10°C$ with a level section at 0°C. The rate of fall of temperature from 0°C to $-10°C$ is quite impressive.

REFERENCE

1. *Handbook of Chemistry and Physics* (42nd edn) (Cleveland, Ohio: Chemical Rubber Publishing Company Ltd,) p. 2309.

Some experiments with radiant heat

E. M. Royds-Jones, Wykeham House School, Fareham

The new school balometers [1] that are now available open up interesting and instructive experiments that previously were too slow or too inconclusive.

1. The most obvious experiment is 'Leslie's cube'. If, instead of a cube, one uses an ordinary tin can painted matt black on one side, the difference in radiation is convincingly shown by the immediate movement of the needle when first one side and then the other faces the balometer. An interesting effect can be observed if part of the can is covered with 'chromium tape'. This is much more highly reflecting for light than is the plain tin, yet it is found to radiate more heat than the tin. A possible explanation is that the reflection is due to metallic flakes embedded in plastic tape which is transparent to light but not to radiant heat of long wavelength.

2. The 'greenhouse effect' can be seen in a way that strikes home to pupils far more thoroughly than blackboard lectures. An electric torch is shone on the balometer and then a piece of glass is interposed. The radiant heat received through the glass is about three-quarters of that received without the glass. But if a hot blackened can (the 'Leslie's cube') is used as the source of heat the radiant heat is completely cut off by the glass. In fact if the glass is put in front of the balometer to start with, then bringing the hot can near makes no change in the reading of the balometer needle. The glass is then removed to show the large reading without it.

3. The cooling of the ground on a clear night to form dew or frost can be illustrated. If the balometer is originally set to read a high figure, a can of iced water placed in front of the instrument immediately makes the reading go down, showing that the balometer is radiating heat out to the can. If a book, or anything at room temperature, is interposed the needle returns to its original reading, illustrating how clouds prevent the cooling of the earth's surface.

4. Liquids and solids can be compared for transparency to radiant heat. With liquids, allowance must be made for the container, the effect of which can be measured when empty. A solution of iodine can be made dark enough to stop all light, and yet it will transmit a considerable amount of radiant heat.

No doubt there are many experiments to show the reflection and refraction of radiant heat, though we have not yet attempted them.

REFERENCE

1. School balometers can be obtained from Grove Industries, Grove House, Grove Road, Fareham, Hants.

Demonstrating a 'solar cannon'

Alan Ward, Saint Mary's College, Cheltenham

According to a news item in the *Daily Telegraph*, 'the average energy from the sun falling on a square metre of British soil throughout the year is about 50 watts' It reminded me of an anecdote in that wonderful old textbook *Nature's Mystic Movements* (Pitman, 1933) by imaginative A. T. McDougall. He mentioned a gentleman who was killed instantly by a shot from a flint-lock pistol, the trigger of which was not pulled. It was neither murder nor suicide. Examination of the room showed that a water-filled goldfish bowl had acted as a lens, focusing the sun's rays on to the pistol's flashpan. Intense heat ignited the gunpowder and discharged the loaded weapon.

In any discussion of how solar energy might be harnessed, children will be fascinated by this bizarre accident. Indeed, I believe the story inspired a 'perfect crime' plotted during a 'Molecule Club' dramatization of scientific principles at London's Mermaid Theatre. Also, a recent photograph in the *Observer* colour magazine showed a solar alarm clock, which was a combination of sundial and a lens pre-focused to discharge a little brass cannon.

These ideas can be illustrated with the aid of a 'flat' 4-ounce medicine bottle, a cork, four red-topped matches, some pencils, and a magnifying glass. The object is to improvise a 'solar cannon'. Wedge the close-packed tail ends of the matches into a hole in the narrow end of the cork. Fix the cork—*not too tightly*—into the bottle. Then set the bottle upon the pencils, which are to serve as rollers. Do this on a bench top in strong sunlight. *Of course safety must be attended to, so observers can be placed behind a transparent screen.*

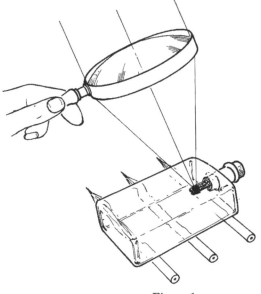

Figure 1

Use the magnifying glass to focus solar radiation through the bottle glass, on to the matches. Eventually a match smoulders, before suddenly bursting into flame, and setting fire to the other matches. The pressure of evolved gases ejects the cork like a bullet—and, at the same instant, the force of reaction drives the bottle gun back on its 'carriage' of rollers.

A hot-water bottle and a (solar) 'serpent'

Alan Ward, College of Saint Paul and Saint Mary, Cheltenham

A safe way to experiment with heat is to use a hot-water bottle. It can be brought into a primary school classroom, wrapped up in a blanket—and discussion can begin at once, about using water for storing energy, and on the value of wool as an insulator. One use for the hot-water bottle is to wrap around a glass bottle that has a balloon stretched over its neck, to show that air expands when it is heated. My most amusing use so far is to provide an up-draught of hot air, to operate a gaily decorated, stiff-paper spiral toy 'serpent' that I suspend from the ceiling, via cotton and BluTak. The model stops rotating when I hold a tray over the heat source, suggesting that it is something coming up from the vicinity of the hot-water bottle that is making the serpent spin. When air in the room is cold, I find that the paper snake rotates slowly above my bared warm arms. So I can get it to work with body heat.

Discussing the serpent with teachers, I was challenged to prove that it was *air* ascending above the hot-water bottle. First I thought that it would be appropriate to demonstrate that an upward breeze would turn the spiral. Using a battery-driven fan, I showed that blowing air upward made the serpent turn clockwise (and then fanning air downwards made it go anticlockwise). Then we sought a way to 'mark' the air with a tracer; we tried banging a dusty chalk-eraser with a stick. Motes of dust by the side of the serpent moved about in random ways, although it was possible to observe definite upward swirling of the chalk motes in the air column above the hot-water bottle. A spotlight might have made this effect more easily visible. The teachers agreed that our 'playing about' was in the tradition of good primary science.

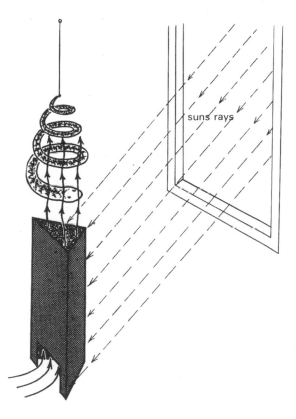

However, I keep going back to that ridiculous toy, now hanging in my study. This morning I wondered if it might be possible to fix a very small d.c. electric motor—*to run in reverse as a generator*—at the serpent's point of suspension, and to connect it to a milliammeter (but I don't think it would work with a hot-water bottle). My other idea, based on something that caught my eye in an American textbook, definitely has possibilities: a 'solar serpent'. I have constructed a triangular tube (three 10 × 50 cm panels), using black paper. There is a little 8 × 8 cm opening in the base of it. This contraption I stand, where sunlight can warm the black paper, just under the serpent. And it works. There is plenty of scope for experimentation here. The theory is that the sun heats the 'solar panel' and warms the air inside the tube, producing a rising convection current to drive the snake. I have not yet tried a black metal tube.

A fresh look at bimetallic strips and their use as thermostats

M. A. Parkin, Western Middle School, Wallsend, Tyne and Wear

Many books concerned with the expansion of solids explain how a bimetallic strip or compound bar is made by riveting or permanently fixing together two strips of metal which have different coefficients of linear expansion. 'The linear expansivity of a material is *defined* as the increase in length when unit length is heated through 1°C.' Table 1 shows the most common metals used in the manufacture of compound bars and bimetallic strips and their corresponding coefficients of linear expansion.

Metal	Linear expansivity $\times 10^4$
Brass	0.19
Copper	0.17
Nickel	0.13
Iron	0.12
Invar	0.009

Table 1

When heat energy is put into a bimetallic strip it bends. A simple explanation of this is, that the metal with the greater linear expansivity increases in length faster and attempts to 'slide over' the metal with the smaller linear expansivity. As the two strips are fixed firmly together 'slide' is impossible, consequently the whole strip bends. As the strip cools to its original temperature the metal with the greater linear expansivity contracts faster than the metal with the smaller linear expansivity and the strip levels out. If heat energy continues to be taken out of the strip it will bend in the opposite direction. The direction of the bend depends on whether the metal with the greater linear expansivity is on the upper or lower surface of the bimetallic strip (see Figure 1). This fact is important if the strip is to be used as a thermostat.

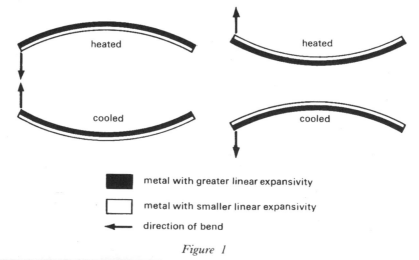

metal with greater linear expansivity

metal with smaller linear expansivity

direction of bend

Figure 1

The work that follows shows how a crude testing device was made which measured the amount of travel (per °C) of a bimetallic strip as it bends, and a further two experiments, one showing the use of a bimetallic strip in a 'heat/cold' alarm system; the other shows how a bimetallic strip can be linked into an electrical circuit to operate a heating/cooling system which compensates for extreme variations in temperature.

MEASURING THE TRAVEL OF A BENDING BIMETALLIC STRIP

The principle of the bimetallic thermostat is for the bending strip to touch a contact point and thus complete an electrical circuit. It is desirable to know how far the strip will move for each degree rise or fall in temperature in order that the contact point can be set at a predetermined distance from the blade of the strip. To determine the amount of travel of the bending strip the following crude but simple and effective device was set up (see Figure 2).

A bimetallic strip (Griffin & George, Cat. No. XHK-740-E) was clamped across a heat resistant board with a boss which was levelled by placing a small piece of metal tube between the board and the screw of the boss. A sheet of drawing paper was Sellotaped to the board immediately behind the bimetallic strip. A small hole was drilled in the board to take the stem of a rotary thermometer. A pencil line was drawn along the length of the bimetallic strip on to the paper, taking care not to cause the blade to dip. A final recheck was made to ensure that the blade was level and immediately in front of the horizontally drawn line. Slight adjustment had to be made. A mark was then made on the paper

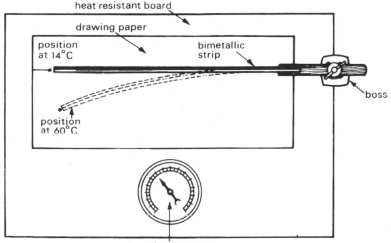

Figure 2

opposite the end of the blade, a note was made of the temperature and the device placed in a drying cabinet. The heat was switched on and when the temperature in the cabinet was at its maximum the door was quickly slid open and a pencil mark was made on the paper opposite the end of the blade as before. A note was made of the temperature and the heat turned off. When the cabinet was cool the paper was removed from the board and a line was drawn between the two points. The length of the line was then measured in millimetres.

The results were as follows: The initial starting temperature was 14 °C. When the strip was heated to 60 °C it moved a distance of 21 mm. Therefore the distance moved per 1 °C was

$$\left[\frac{21}{60-14}\right]$$

0.456 mm. The strip was tested a further three times and a final average figure of 0.5 mm/°C was accepted.

It is important to note that a straight line only was drawn between the two points as the crudity of the device did not merit attempts to measure or calculate the length of the arc described by the movement of the blade of the bimetallic strip. (Refinements are possible however.)

THE BIMETALLIC STRIP USED IN A 'HOT/COLD' ALARM SYSTEM

The experiment is set up as in Figure 3. The buzzer (or bell) and lamp are fixed to a wooden board. The bimetallic strip is held horizontally with a boss fixed to a retort stand. A crocodile clip with lead (plug opposite end) is attached to the blade, near the brass handle. The plug is placed in the 'O' socket of the transformer or either terminal of

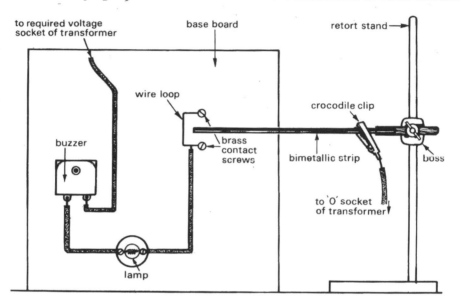

Figure 3

a low voltage supply. The height of the bimetallic strip is adjusted to be equidistant between the two brass terminals or screws in the board. On deciding the distance apart for the screws, refer to the previous section on the measurement of travel of the bending strip. The distance between the screws will depend on how many degrees above or below room temperature that the alarm system is to operate. The two screws are linked together by a loop of copper wire. A wire is attached to this loop and the other end connected to one terminal of the lamp. A short wire connects the remaining terminal of the lamp to one terminal of the buzzer and finally a wire from the other terminal of the buzzer is connected by plug to the required voltage socket of the transformer or the other terminal of the low voltage supply. (It can be seen that the lamp and buzzer are linked in series.)

To operate the system as a 'heat' alarm it is necessary to increase the temperature around the bimetallic strip. One way of achieving this is to place a lighted bunsen burner near to, but not necessarily touching the strip. Depending on the sensitivity of the strip or the closeness of the contacts, extra heat can be produced by the flame of a candle placed below the strip. As the strip bends downwards it touches the lower contact screw, the circuit is completed and the buzzer sounds, simultaneously the light illuminates indicating that the temperature of the room or surroundings has risen above the set temperature.

Allow the strip to cool to room temperature by removing the heat source, then carefully spray the area around the strip with 'Electrolube' aerosol freezer spray; first spraying the children's hands to show its cooling effect. The bimetallic strip begins to curve upwards and eventually makes contact with the upper brass screw, once again completing the circuit which operates the buzzer and lamp, indicating that the temperature has fallen below the required room temperature. When spraying ceases the strip levels out. The contact is broken and the alarms cut out.

THE BIMETALLIC STRIP AS A CONTROL DEVICE

This experiment shows how the bimetallic strip can be used to compensate for variations in temperature of a room. The apparatus is set up (see Figure 4). A small heating coil made from 26 swg constantan wire represents an electric fire and a small, low voltage electric motor with a homemade fan blade attached represents a cooling fan. These two are wired in parallel into the circuit. (Note that the contact points are not joined by a wire loop in this experiment.)

If extreme conditions are created as in the last experiment the appropriate agency comes into operation to compensate for the extreme. When the temperature rises above the required temperature the bimetallic strip bends downwards and touches the lower contact point—this completes the circuit for the fan—which operates in an attempt to lower the temperature. When the cold conditions are simulated with the freezer spray the bimetallic strip bends upwards and makes contact with the upper contact point, which completes the circuit for the heating coil which operates and glows red, while the temperature of the room is below the required temperature.

At this stage it is important to mention to the children involved, that as the system in Figure 4 is 'open', then the fire and fan will not be switched on selectively by the bimetallic strip as the small draught produced by the fan is not powerful enough to cause the strip to move, neither is the current directed to the vicinity of the strip. The heat produced by the coil is not sufficient to cause movement of the strip, even though the contact points may be set close together to give maximum sensitivity. As a refinement, however, the system could be enclosed in a shallow box with a perspex front (for viewing) and two ventilation holes, the inlet near to the fan and the outlet just above the bimetallic strip, consequently a cooler air current could be produced which would flow

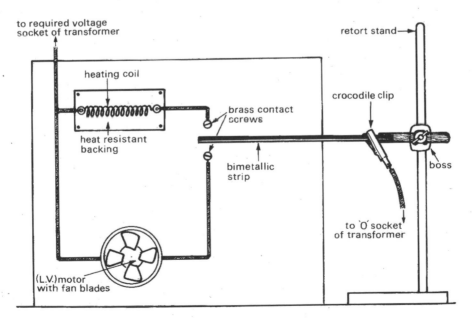

Figure 4

around the strip and cause it to move. In turn the heat from the coil would be retained in the box and this 'confined' heat would activate the strip. With the 'open' systems extreme conditions have been induced using a bunsen burner and freezer spray but if the principles involved were incorporated in real temperature control situations then the natural changes in temperature would operate the bimetallic strip (thermostat). If for example the room became warmer, due to the sun's rays shining through a window, or cooler because of the usual coolness of the evening, then the respective cooling or heating system would come into operation.

The main advantage of this work is that children are able to follow an investigation through from the basic idea that a bimetallic strip bends when it is heated, to the design of an apparatus to show its usefulness. The children can actually see the bimetallic strips working in response to temperatures above and below the ambient temperature of the room. In many cases, thermostats working on mains voltage are either enclosed or sealed for obvious safety reasons with the exception of some aquarium heater/thermostats which are sealed but can be seen to operate through the glass tube casing, consequently, to be able to see safely the movement of the strips gives greater understanding and meaning as to their use in temperature control situations as the last experiment showed.

To reinforce this work on bimetallic strips it is useful to make a collection of old obsolete or broken thermostats for comparison purposes and to show that there are many different ways of incorporating the basic idea of the bimetallic strip into thermostats.

FOOTNOTE

Invar, is the trade name for an alloy of iron, nickel and carbon and it has a very small linear expansivity.

Calibrating a thermometer's upper fixed point

Ron Cox

What a delight it is when a young pupil (Jonathan Gemain – third year) spots an obvious improvement on an age - old experiment!

RV Cox, Gresham's School, Holt, Norfolk NR25 6EA.

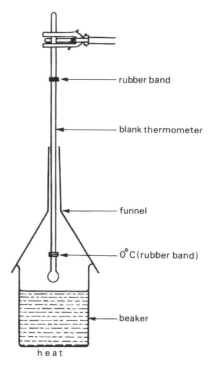

rubber band

blank thermometer

funnel

0°C (rubber band)

beaker

heat

Investigating water heaters with the least able

M David-Tooze and REL Waddling

Investigating the use of electricity to provide hot water in the home is a key issue in many GCSE science courses. Our least able students follow a modular Applied Science scheme (mode 2 SEG) making use of the Science at Work booklet - *Domestic Electricity* [1]. Some of the very weakest groups have found difficulty with the approach adopted and the purpose of this note is to suggest alternative strategies.

AN ALTERNATIVE FILAMENT

It is suggested that a filament can be constructed from lacquered nichrome wire. If uninsulated wire is used there are short circuits across the filament. As an alternative we recommend that a filament is constructed by passing about 50 cm of uninsulated nichrome wire through a glass tube (OD 4 mm and fire polished) and then winding a coil back around the

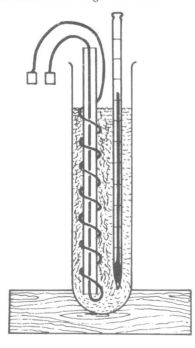

Figure 1   Diagram of heater

Investigating the effect of different thicknesses of wire
It is important that students appreciate the need for controls. In this investigation the current, volume of water, time of heating and the initial temperature of the water need to be kept constant.

Immediately we are faced with the difficulty of using ammeters. Not only do most have two scales which can confuse, but they can also be very easily damaged by accidental overloading. To overcome these problems we have used a 24 W (12 V) bulb in place of the ammeter.

The procedure that we have employed is for the teacher to demonstrate the construction of the filament and the setting-up of the circuit using 22 swg wire. The teacher inserts an ammeter into the circuit in series with the bulb, to show that the brightness of the lamp depends upon the current flowing in the circuit. The rheostat can then be adjusted to give '1 amp brightness' and the bulb left on as a standard (see Figure 2).

Students are then able to construct their own filaments and circuits and adjust the brightness of their bulbs to give a current of approximately 1 amp. The water heaters should be left on for

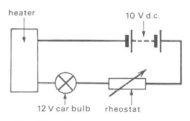

heater   10 V d.c.

12 V car bulb   rheostat

Figure 2   Circuit

ten minutes. We extend the practical by asking students to record the temperature every minute and to then plot a line graph of their results. This procedure is then repeated using wires of different thicknesses. After this time a temperature rise of approximately 20 °C will be recorded for 34 swg wire and of approximately 6 °C for 22 swg wire. These are much greater temperature rises than we had found when using the method in the booklet.

While the classes were performing this investigation we inserted an ammeter into each group's circuit to see how accurately the students had been able to adjust the current. Nearly all students were within 0.1 A. This was sufficiently accurate to show the expected difference between fairly similar wires, for example 32 and 34 swg.

Investigating the effect of using different currents
A similar approach was used with students using 32 or 34 swg nichrome wire. Different current can be passed for the same duration of time. Again graphs can be plotted. The teacher must demonstrate and leave on bulbs of different brightness. The filament should just glow for '0.6 A brightness' and for '1 A brightness' it will be dimly lit. All students will be able to investigate two different currents, while the more able students are able to investigate intermediate values. Typical results obtained after ten minutes of heating are shown in Table 1.

THE EFFECT OF LAGGING

The range of suggested experiments can be extended by this investigation, which can also be

Table 1   Temperature rise (°C) after ten minutes

current (A)	swg		
	22	30	34
0.6	1	6	8
1.0	6	19	21

used as a problem-solving exercise. Students can wrap the test tube in different lagging materials in turn and measure the temperature rise after ten minutes (using 32 swg wire). The rate of cooling of the hot water could also be investigated.

REFERENCE

1 Science at Work, *Domestic Electricity*, (Addison Wesley, 1979).

M David-Tooze and REL Waddling, Falmouth School, Cornwall

A simple latent heat demonstration

Albert Wood

Recently, it has been possible to buy hand warmers which operate by releasing latent heat as a liquid solidifies. These take the form of a transparent plastic bag containing a clear, colourless liquid together with a metal disc (Figure 1). On

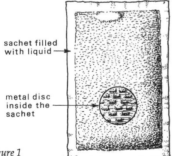

Figure 1

close examination the metal disc is found to be spherical. Also, its surface has raised markings. If pressure is applied to the disc so as to reverse its curvature, the liquid in the bag solidifies and the bag becomes warm.

When a colleague bought one of these hand warmers, I was so impressed that I asked some boys to investigate its construction. It soon became clear that it was possible to produce a class set of simple apparatus to give a memorable demonstration of latent heat in action.

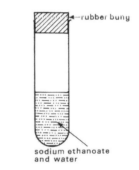

Figure 2

Our liquid was 20 parts of sodium ethanoate (trihydrate) to 3 parts of water, by mass. The 'hand warmer' was a boiling tube one third full of this liquid, with a rubber bung to act as a seal (Figure 2). The device was heated in a water bath and gently agitated until a clear solution of sodium ethanoate had been formed. It was then left to cool in the air, or cooled more rapidly in cold water. The method of cooling was not critical, provided that the solution was not shaken up and that no undissolved crystals were allowed to make contact with the solution.

When the boiling tube was cool to touch, it was shaken up so that the solution inside made contact with the rubber bung. This contact triggered crystallization and the whole device became warm. The apparatus is obviously simple and inexpensive. In addition, it can be used many times over without having to adjust the solution inside.

It was interesting that the rubber bung triggered the crystallization, whereas many metals seemed to have no effect. A notable exception was stainless steel. A supercooled solution of sodium ethanoate began to crystallize as soon as it was touched with a stainless steel spatula, even though the spatula was clean and had a smooth surface.

A similar hand-warming effect could be demonstrated using sodium thiosulphate (5 hydrate). This chemical did not require the addition of any extra water, but the effect was much less good.

⚠ SAFETY

The chemicals used do not have any particular hazard associated with them. Standard precautions to protect eyes and clothing are probably sufficient. In addition, it should be noted that the cold solution of sodium ethanoate will crystallize, and therefore release heat energy, if it makes contact with skin. Even a drop of it landing on the wrist could be quite painful.

ACKNOWLEDGEMENT

Thanks are due to Stephen Lewis and Andrew Tierney, who conducted many experiments to perfect the hand-warmer.

A Wood, Whitgift School, South Croydon

Pinhole camera without a dark room

J. M. Wilson, Government Secondary School, Kano, Nigeria

A simpler version of the pinhole camera as described by R. L. Jones (*S.S.R.*, 1963, 154, **44**, 731) can be made by any pupil in a few minutes from a tin (with lid), a piece of tissue paper, a newspaper and a piece of string. A small hole is made in the bottom of the tin with a nail. The lid is removed and replaced by the tissue paper tied with string. The folded newspaper is rolled round the tin with one edge close to the tissue paper and the other edge pressed against the face, as shown.

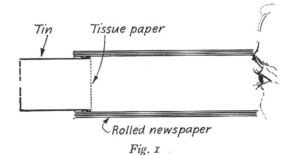

Fig. 1

A 'dark room' for a standard pinhole camera may be made from an open cardboard box. A flap is cut in one side into which the camera fits (if the flap is left in place it helps to seal the 'dark room'). A peephole is cut in the opposite side. Pupils must realize that the peephole through which they look is not the 'pinhole'.

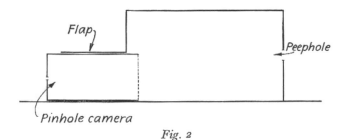

Fig. 2

Detection of infra-red using a phototransistor

T. J. Ericson, St Helens Technical College, Lancs.

A narrow parallel beam of light is split up by a glass prism and the spectrum shown on a sheet of white paper. The phototransistor OCP71 is moved across the spectrum and the output indicated on an oscilloscope.

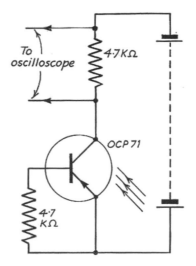

If no d.c. amplification is available on the oscilloscope, the light beam can be chopped with a rotary shutter. With this modulation of the light beam the experiment can be performed without any special darkening of the classroom.

The circuit diagram for the phototransistor is as shown. The oscilloscope used was Telequipment S51E.

Total internal reflection

George Cousins, Morpeth Girls' Grammar School, Northumberland

This is an adaptation of the 'empty test-tube' experiment for showing total internal reflection. The empty test-tube is normally placed in the water at an angle of more than 45°, incident light as shown, and when viewed from above the glass has a silvery

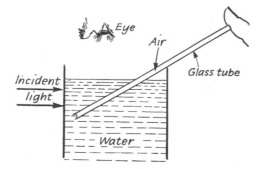

appearance. On introducing water into the tube the silvery surface disappears, but this operation is time consuming, and the experiment loses impact.

Replacing the test-tube with a clean piece of glass tubing, the air is trapped in the tube by closing the end with the index finger before immersing it in the water. The silver appearance goes when the finger is removed and the water flows into the tube.

Introducing fibre optics

Alan Ward, St Mary's College, Cheltenham

A crooked solid glass rod can be used to demonstrate the principle of the 'light pipe' by means of which much of the light entering one end is transmitted 'around corners' by total internal reflection, to emerge at the opposite end. The rod is easily improvised with the help of a glass-cutting knife and a bunsen burner.

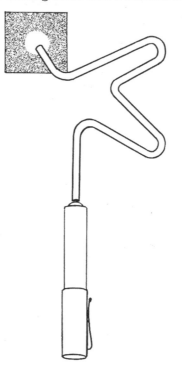

Light from a pen-torch shone into one end of the rod can be seen at the far end directly, or indirectly by reflection on a small paper 'screen' held near the exit (see the illustration). But some of the light escapes; so if the illuminated rod is viewed in darkness, its full length seems to glow, which makes a very pretty sight.

Another light pipe is produced by tying a loose knot in transparent plastic tubing that is afterwards filled with water. Air bubbles must be excluded. Then, with two clamps to hold the tube ends, the tubing can be mounted on a retort stand. Some of the torchlight shone into one end of the tube can be seen at the other end, after travelling around in the knot.

It is also interesting to shine a torch into a handful of glass fibre, in a dark room. Simple experiences like these can lead to a consideration of fibre optics, the technology of transmitting illuminated 'pictures' through flexible cables consisting of thousands of separate glass fibres, each of which can carry a little 'bit' of the picture.

How to spear a fish—a simple illustration of the effects of refraction

Alan Ward, St Mary's College, Cheltenham

To spear a fish in water from the lakeside, the weapon must be aimed towards a place somewhat closer than the fish appears to be. This is because light reflected off the fish is bent towards the surface ('away from the normal') when it comes from water into air. Eyes and brain do not sense the bending, so the light seems to come from a place under water higher and further back than the real location of the fish.

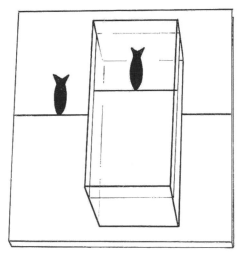

The point can be illustrated by an easy controlled experiment. Paste a pair of identical black paper 'fishes', about three inches apart, at right angles to and just touching a straight line ruled across a white sheet of paper (see the illustration). Then stand a narrow transparent plastic lunch-box filled with water over one of the fish, and compare the fishes' real and apparent locations.

Colour mixing by shadows

C. J. Feetenby, Allerton High School, Leeds

When the conventional experiments on the mixing of coloured lights have been performed, a striking demonstration can be shown by placing an obstacle, e.g. a hand, between the lamps and screen to cast shadows. The diagram shows how three such shadows are cast giving magenta when the green light is cut off, cyan when the red light is cut off and yellow when the blue light is cut off.

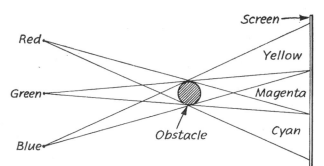

An improved model eye

M. H. J. Hawkins, Bexley Grammar School

The eyeball is a goldfish bowl (Figs 1 and 3) filled with dilute fluorescein solution which acts as both vitreous and aqueous humours. A lens which can accommodate is made flexible by facing it with polythene sheet. The difficulty of making it converging is solved by decreasing the air pressure inside so that it becomes biconcave. Ample control is obtained by moving the reservoir of the mercury manometer about 15 cm.

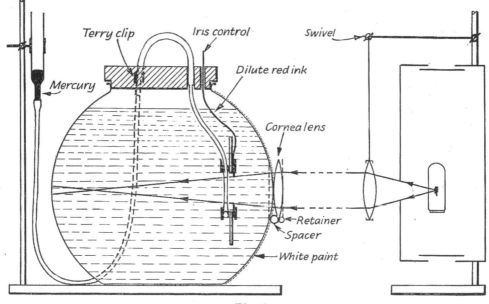

Fig. 1

The method of construction can be varied to suit the materials and facilities available. The one illustrated (Fig. 2b) avoids the use of special tools, so it can be recommended to any handyman. The galvanized wire was bent round a cylinder of suitable diameter and cut to leave a gap for the connecting tube. The plastic tube (8 mm o.d.) was prepared, cutting its ends radially to the final circle, and eased on to the wire

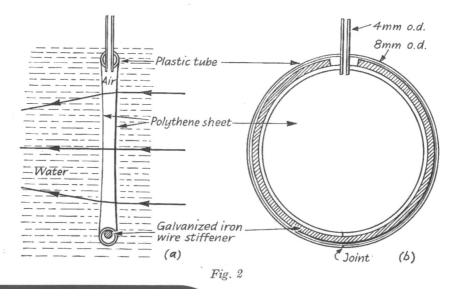

Fig. 2

until the ends met opposite the gap. The joint was sealed with Bostik 1 and allowed to harden before using a cork-borer to make the hole for the connecting tube. This tube was fitted next and the joint also sealed with Bostik 1. The lens faces were attached with the same adhesive, placing each side in turn on to a taut sheet cut from a clear polythene bag. It is important that the circle should be true and flat to avoid astigmatism later. The completed lens was clamped to the back of a simple four-leaf iris, but a plate with a circular aperture would serve quite well.

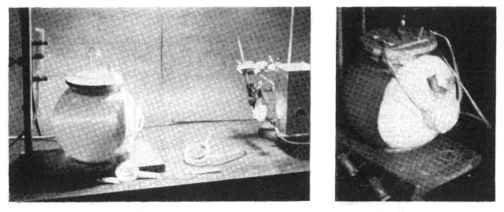

Fig. 3a   Fig. 3b

The cornea must be modified to demonstrate long, normal and short sight, so a holder similar in construction to the lens holder was stuck on to the outside of the bowl. Another stiffened semicircle was first fixed as a spacer, and to this a slightly smaller U, made of 4-mm o.d. plastic tubing, was attached as a retainer. The front of the bowl was then painted white to shield the 'retina' from stray light during near-point demonstrations. The photos of Fig. 3 show the frame which takes the interchangeable correcting lenses. The labelling tags provide convenient handles for the lenses.

A car headlamp can be used as a point source, but a quartz–iodine lamp is much better. To facilitate a smooth demonstration it is worth while arranging a quick change from distant to near object by swivelling the lens which gives the parallel beam in and out of position. The lamp alone can be placed at the near-point to suit the model.

There are likely to be several variables which will affect its performance, but the following component sizes may serve as useful guides:

Cornea–lens:	long sight	$f = 50$ cm
	normal sight	$f = 20$ cm
	short sight	$f = 10$ cm
Correcting lens:	long sight	$f = 50$ cm
	short sight	$f = -30$ cm

I am very indebted to Mr H. Keely for his unfailing help in trying many designs before a satisfactory one was produced, to C. D. Parrott who undertook the construction of the iris and to Mr E. W. Heasman for his ready co-operation.

Section 3: Optics

Seeing through a brick

J. C. Siddons, Thornton Grammar School, Bradford

The arrangement of mirrors whereby one can 'see through a brick' is well known. A smaller version can be made in a boot-box as shown in the figure. The strips of mirror can be held in place by slotted pieces of wood. In place of the brick, black paper will

---

* Very kindly supplied by the Morganite Carbon Co., Romford, Essex.

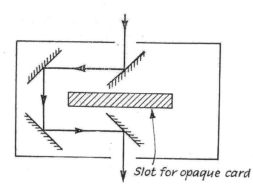

Slot for opaque card

serve as the object to be 'seen through'. If possible aluminized strips of mirror should be used to cut down the multiplication of images.

The box can be used in two ways. For individual use it can be held up to the eye; for class demonstration a beam of light can be shone through it. Ask a student to look through the box at one of his fingers held near the entrance hole and to comment on anything different about it. The finger will look smaller than expected, for in effect it is further away.

Image and object

J. C. Siddons, Thornton Grammar School, Bradford

Every schoolchild knows what happens to words and numbers when they are held up in front of a mirror: they are 'turned round'. (We come back later to some apparent exceptions.) This turning round however can be brought about simply by writing on thin paper or, better, glass and then looking at the words from behind. Fig. 1 shows both methods of turning writing round at work. Words on paper are two-dimensional objects: with such objects the difference between them and their mirror image is simply one of point of view.

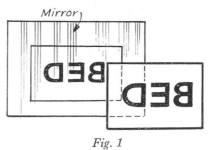

Fig. 1

If three-dimensional objects without an axis of symmetry are used, then there is a much bigger difference. To show this hold your left hand in front of a mirror with its palm facing the mirror and your right hand to the side of the mirror with its palm facing you (Fig. 2). The image of the left hand will be seen to be a right hand: no changing of point of view will remove this difference.

Besides using hands, use wire helices: make two helices of equal size but wound in opposing directions. Either one held in front of the mirror becomes the same kind of helix as the other held alongside the mirror. If two model cars were used, one an English type and the other its Continental counterpart, then the mirror will turn one

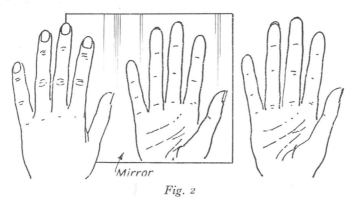

Fig. 2

into the other. Then there is the human face, in particular one's own. Hold a photograph of your face alongside the mirror and compare the face in the photo with the face in the mirror. (If you do not have a big photo of your face, diminish the mirror image by using a convex mirror.)

If instead of one mirror two mirrors are used to produce the final image the latter will be the image of an image. The first reflection will make right left, the second reflection will make left right and so the final image will be the correct way round, i.e. it will be identical in shape with the object. Set up two mirrors in wooden slots so that they are exactly at right angles and touching along a vertical line. Aluminized mirrors are preferable for then there is no break in continuity where the mirrors meet. For objects use as before your hands, wire helices, toy cars—but especially one's face. Burns' gift is then given us; we see ourselves as others see us. If you happen to have a symmetrical face, make it unsymmetrical by winking.

Let us return now to two-dimensional words. An amusing trick can be played—if it is considered educationally advisable—which depends upon the fact that some capital letters have a horizontal axis of symmetry. These letters are B D E H I O X and if hand-printed C and K: along with them are the digits 3 8 and 0. With a red pencil write a word made up of these letters, e.g. HIDE: with a blue pencil write a

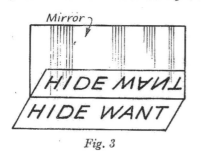

Fig. 3

word made up of letters not in the list, e.g. WANT. Place the words in front of a mirror as shown in Fig. 3. The word in blue is turned upside down, the red one apparently not. The gullible observer can be persuaded that red light is reflected differently from blue light.

Capital letters with a vertical axis of symmetry are AHIMOTUVWXY. A palindrome is a word which reads the same either way, e.g. radar. Any palindrome made of these letters would look the same, i.e. not turned round, held up in front of a mirror. I can find none longer than TUT-TUT or TIT. In addition a word such as ATOM if it is printed with the letters in a vertical column appears unchanged by the mirror. Thus the previous trick can be repeated; a vertical red ATOM contrasts well with a vertical blue BELL.

Pepper's ghost*

J. C. Siddons, Thornton Grammar School, Bradford

The figure shows the arrangement as seen from above. A slip of card A behind a glass sheet S bears the letter A which faces S. A second card bears the letter B,

suitably inverted, which also faces S but from in front. L is a lamp, e.g. 12 V, 24 W which illuminates card B. A and B are so arranged that the image of card B is alongside the card A. It helps if two screens M and N are suitably placed, M to prevent too much light falling on A, N to shield the eye of the pupil who is standing on the B side of S.

He is told to look through S and observe the two letters: next he is asked to put his hand behind S and pick up A—this he can do. Then he is asked to pick up B which to his consternation he cannot do. Pepper's ghost still startles boys and girls, if not their teachers.

* Demonstrated in the ASE Members' Exhibition at Stirling.

An optical illusion with a spotting tile

Alan Ward, Saint Mary's College, Cheltenham

When moon photographs are inverted, concave craters sometimes look like convex mounds. The well-known optical illusion is discussed by pioneer lighting engineer Luckeish in his famous book *Visual Illusions* (1922)—now available in paperback from Dover Publications, Inc. The familiar explanation is that the brain assumes that what we see is generally illuminated from above. Light coming from below eye level casts shadows which baffle the brain into making dramatic errors of judgement. Recently the sight of an opaque 113 mm × 95 mm polystyrene spotting tile with twelve concave depressions (Philip Harris C2017) suggested an experiment.

I taped the white tile 65 cm above bench level on the wall of a darkened room. Then I moved a 60 W 'frosted' light bulb back and forth on the bench, while watching to see if the illusion would happen. Suddenly the tile cavities actually did resemble mounds—and it was virtually impossible to perceive them otherwise. The 'humps' had a waxy transparent appearance. This occurred when the light was 75 cm away from the wall (between my eyes and the tile). Colleagues and students confirmed the observation. We also noticed that convex humps looked concave when the tile was reversed. But cavities can be made to appear as humps in a daylit room or outdoors, if the tile is held in the hand and tilted certain ways.

A pinhole camera kit

P. Gee, Dr Challoner's Grammar School, Amersham

We have been using the standard cardboard box with holes and removable lid. These do not last long, and in any case the lads regard them as special pieces of apparatus.

Superior kits can be made from standard plastic drainpipe. Cut, with hacksaw, into suitable (15 cm) lengths, tear some aluminium cooking foil and greaseproof paper into squares and find some elastic bands which just go round the pipe.

Pupils then press the foil over one end, the paper over the other, and hold these on with the bands. Make one, two . . . n pinholes, and finally finger hole in the foil. It is true that they cannot put the lens between hole and screen but we never found this part very convincing anyway.

The straw electroscope

Lalit Kishore, Mayo College, Ajmer, India

An electroscope can be made very simply as follows, and it works wonderfully. Take a cardboard box of the length greater than that of the drinking straw. Cut off the front and back face of the box except the two strips as shown below. Make a cut

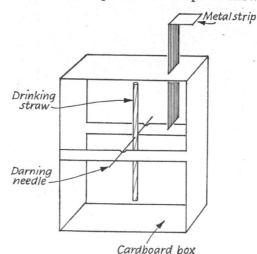

Metal strip

Drinking straw

Darning needle

Cardboard box

on each of the strips. Pass a darning needle through the straw about 0.2 cm away from its centre. Through the top face of the box pass a metal strip as shown. The electroscope is ready for use now. On bringing a charged body near the metal strip the straw shows a deflection. Furthermore, the electroscope can also be charged. Better results are obtained if the straw is painted with 'silver' paint.

Measuring the Muller–Lyer optical illusion

Alan Ward, Saint Mary's College, Cheltenham

Draw two identical straight lines with arrowheads on both their ends. But on one line draw the arrowheads inside out (Figure 1). Now one of the equal lines looks longer than the other. This is of course the famous Muller–Lyer optical illusion, to be found in most puzzle books for children. Can the 'degree of error' be measured?

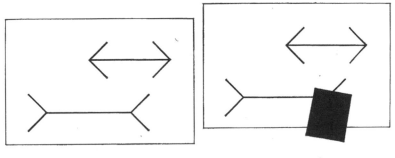

Figure 1   Figure 2

We began with an enormous diagram in which the straight lines to be compared were each 45 centimetres long. We also used a 15 cm × 20 cm black card and a ruler. Then we asked each member of the class, in turn, to cover the 'longest-looking' line with the card—so that it appeared equal in length with the other one (Figure 2).

Measurements of the uncovered portion of the line were recorded on a bar graph (Figure 3). In a group of thirty Saint Mary's students, no less than seven individuals left only 40 centimetres of the line uncovered. Four students claimed that both lines looked equal. The overall error was about 9 per cent.

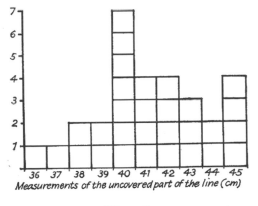

Measurements of the uncovered part of the line (cm)

Figure 3

The activity can be presented conveniently as a 'side-show' during a routine double teaching period. More information on the Muller–Lyer illusion is to be found in *Eye and Brain*, by R. L. Gregory (World University Library, 1966), and ideas for other experiments will be suggested by S. Tolansky's *Optical Illusions* (Pergamon, 1964).

A thumb-controlled 'light pipe'

Alan Ward, Saint Mary's College, Cheltenham

Here is an easy way to make light travel around a corner. Begin by using a warmed cork-borer to make an 8 mm diameter hole (A) near the bottom of a screw-capped plastic squash bottle. Bore a 5 mm hole in the other 'side', near the top of the bottle (B). Cover the bottom hole with the left thumb (if right-handed), while filling the bottle with tap water. Replace the cap, before quickly transferring the thumb to the top hole—thus preventing much water from escaping. Do all this in a dimly lit room.

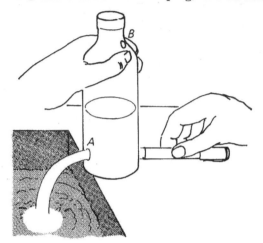

Support the bottle upon the edge of a sink, while using the free hand to shine a pen torch straight through the water, towards the inside of the still-covered bottom hole. Notice that the sink is in shadow. Then uncover the top hole, to cause a jet of water to curve down into the sink. Most of the light entering the jet is repeatedly totally internally reflected, producing a brilliant patch of illumination 'around a corner', where the water hits the sink. As long as the water lasts, the jet can be stopped and restarted.

Demonstrating a solar paradox

Alan Ward, Saint Mary's College, Cheltenham

If a bullet, shot from an imaginary super-gun aimed at the rising or setting sun, could travel instantaneously, it would miss its target by millions of miles. For the sun is actually below the horizon at these times, and, if the earth had no atmosphere, the sun would be below an observer's 'line of sight', and therefore invisible.

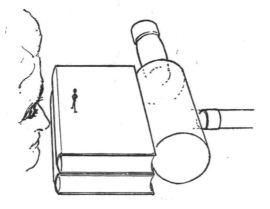

Sunlight reaching the top of the atmosphere from just below the horizon is refracted, or 'bent', slightly downwards, when it comes from space into the relatively denser air. As it enters denser and denser air, it is bent more and more. Consequently it can reach the eyes of an observer (whose brain 'sees' the sun as if it were above the horizon).

Demonstrate this paradox by putting a clear glass bottle of water (the 'atmosphere') on its side, next to a pile of books 4 or 5 centimetres shorter than the diameter of the bottle. Stand a miniature toy 'observer' on the books (see the illustration)—and put a lit pen torch 'sun' on the other side of the bottle. Look towards the bottle, across the top book (the 'horizon').

Although the torch is below a direct line of sight, it is clearly visible through the bottle. Glass and water bend light coming from below eye level, towards the toy figure. But the torchlight seems to be coming from above. Study the effect in a darkened room. The 'skyline' glows with refracted light. Take away the bottle then—and our observer is plunged into 'night'.

A quick measure of persistence of vision

G. Devlin, St Aloysius College, Glasgow

It was while watching his daughter making figures of eight with a sparkler last Guy Fawkes' night that the author thought of a very simple method for measuring persistence of vision. All that is required is that one person A, holding a lighted torch bulb in a semi-darkened room, swings his arm around in a vertical circle. An observer B, standing at right angles to the plane of the circle, watches as A gradually increases the frequency of rotation until he can just see a closed loop. At this point B informs A, who tries to maintain this frequency, while B then measures the time for, say, 10 rotations. One-tenth of this time represents an upper limit for persistence of vision.

The double beam oscilloscope: sound and waves in elementary physics
W. K. Mace, King Edward VII School, Sheffield

INTRODUCTION TO THE OSCILLOSCOPE

Children work in pairs, each pair having a pencil (or better, a felt pen) and a long strip
of paper such as could be provided from a roll of decorator's lining paper sawn in half.
One child executes 'vibrations' by moving the pen rapidly to and fro across the paper
whilst the other draws the paper steadily along the bench. By asking them to vibrate
rapidly, slowly, widely, jerkily, etc., they can observe how the 'shape in time' of
vibration can be made visible as a shape in space; regularity, frequency, amplitude and
'waveform' can be identified in terms of their characteristics in the trace.

The next step is to connect both a vibrator unit and the CRO to the same AFO
using the lowest available frequency and a very slow time-base: they can see their 'pen
traces' being precisely drawn across the screen. By increasing frequency and sweep
speed until persistence of vision takes over, they can be shown exactly what the
oscilloscope is doing.

Having added to the above an extra emphasis on the fact that 'right means later', a
simple and accurate way of measuring the speed of sound becomes possible.

THE SPEED OF SOUND

As illustrated in Figure 1, microphones A and B, connected respectively to Y_2 and
Y_1, pick up a 10 kHz signal from a speaker. A is kept stationary and the oscilloscope
is set to trigger on A. If B is now moved away from the source, the Y_1 trace moves to

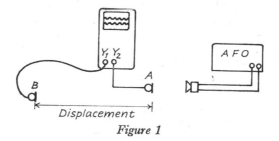

Figure 1

the right as the vibrations arrive increasingly late at B. By counting the number of
peaks passing a fixed point on the screen one can tell how many ten-thousandths of a
second late the signal is. At about 35 cm it is ten vibrations late, which amounts to
$\frac{1}{1000}$ s, giving a speed of 350 m/s.

SUPERPOSITION OF WAVES

This is frankly a *simulation* demonstration, and this must be firmly emphasized. It is
intended as a reinforcement or clarification after everything possible has been shown
with real waves.

The CRO is externally triggered, conveniently using the 2 V output from a mains
transformer unit. If then a signal at 49 Hz is applied to Y the trace will move slowly
forwards at the rate of one 'wave' per second. 51 Hz will produce a wave moving
backwards. The same effects are obtained using, for example, frequencies each side of
200 Hz. Given two AFOs (one of which must not be earthed), Y_1 and Y_2 can display
any two moving waves, either indistinguishable in wavelength, or in a simple ratio to

one another. Further, by arranging the switching system shown in Figure 2 the sum of these waves can be displayed on one trace; thus one can study in simulation the spatial relationship of two moving wave trains and the result of their superposition.

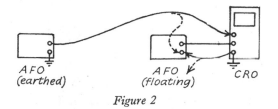

Figure 2

A good thing to show first, to make it clear what the device is doing, is a high frequency of small amplitude superposed on a large amplitude low frequency, the waves moving in the same, and also in the opposite directions. Figure 3 shows the result of using $50 - 1$ Hz and $450 + 9$ Hz.

Figure 3

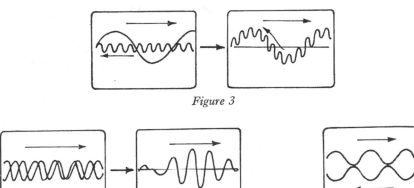

Figure 4   *Figure 5*

Beats can be shown using, for example, frequencies close to 250 and 300 Hz; these can be held steady or made to move slowly across the screen, first separately, then superposed (Figure 4). The formation of standing waves can be looked at very fruitfully by causing two waves of (sensibly) equal wavelength to move across the screen very slowly in opposite directions either on the same line or one immediately above the other (Figure 5). By this means it will be seen that crests coincide only at certain fixed points in space, and the resultant slow standing wave is quite beautiful to watch.

For more advanced pupils it is instructive to look at the effect of unequal amplitudes on beat and standing wave patterns, especially the latter, which is a curious blend of standing and progressive wave.

How big need a mirror be?

Alan Ward, Saint Mary's College, Cheltenham

Usually, when I ask a student how tall a mirror must be to reflect her full standing height, she replies that it depends how far away from the mirror she is standing. The answer is based upon common experience. When she is looking into a small wall-mirror, people a long way behind her appear whole, while only a portion of herself is reflected into her eyes. The surprising truth is worth investigating. So here is a work-sheet for children, followed by an idea for a 'proof' which the teacher can present as a paper-cutting trick. The children should be familiar with the laws of reflection and the properties of images in plane mirrors. I title the work-sheet *Seeing All*.

Hole

Figure 1

MATERIALS

Plane mirror (about 20 cm × 30 cm) fixed to the wall
'Grease' pencil (for writing upon glass)
Tape measure
Stiff card about 25 cm × 35 cm
Cut-out of a standing man or woman about 25 cm tall
Paste

EXPERIMENTS

1. Stand facing the mirror in such a way that you can see the top of your head as if it were touching the mirror's top edge. Then, by pointing a finger at yourself, show an assistant the lowest part of you that you can see reflected. Ask the assistant to take the tape measure, to find the proportion of your body that is being reflected by the mirror. She can do this by measuring from the top of your head to where you are pointing. Repeat the measurement several times while you are standing at different distances in front of the mirror.

How tall must a plane mirror be for you to see the whole of your reflection while you are standing up straight with hands by your sides? *Remember that your measurements will be rather rough.*

2. Paste the picture of a standing figure on to the stiff card. When the paste is dry, bore a neat hole between the eyes of the picture (you can use a cork-borer). Face the picture towards the mirror while you look through the hole at the picture's reflection. Ask your assistant to mark with grease pencil the points upon the mirror at which the top and bottom parts of the picture appear. (You have to tell your assistant where the marks must go.) Measure the heights of both the picture and its reflection. Repeat several times. See Figure 1.

Results should confirm your impressions from Experiment 1.

DEMONSTRATING A 'PROOF' BY CUTTING PAPER

Both investigations lead to the conclusion that a mirror half as high (and wide) as the viewer will do. An interesting final proof can be demonstrated by the teacher, if certain facts about reflection in plane mirrors can be taken for granted. We need to know that angles of incidence equal angles of reflection, and that the apparent size and distance from the mirror of an image is the same as the actual size and mirror distance of the object.

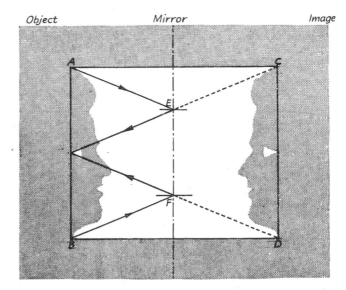

Object   Mirror   Image

Figure 2. The paper can be lightly marked in pencil beforehand

Figure 3

Fold a very big rectangle of drawing paper in half longways, before cutting out the portion shown in Figure 2. The top and bottom points of the profiles must form corners ABCD of a rectangle when the paper is opened out. Fix the 'diagram' to a blackboard, using pieces of adhesive 'putty' (like Plas-Fix, from Philip and Tacey Ltd, Andover, Hampshire). Chalk in the 'mirror' between the creases of the paper—and draw incident, reflected rays and normals, as suggested in Figure 3. The 'object'-observer's eye must be upon the line AB.

Figure 3 shows how the eye sees point A at C, and point B at D. During the demonstration, all assumptions about the diagram's structure must be discussed with the class. Finally, it is only necessary to measure the proportion of mirror between the normals E and F—and to compare this measurement with either the height of the object AB, or the image height CD. It will always be found that EF is exactly half of AB or CD.

ACKNOWLEDGEMENT

The activities suggested on the work card are adapted from ideas presented by Victor Schmidt and Verne Rockcastle, in their excellent elementary science source book, *Teaching Science With Everyday Things* (McGraw-Hill).

A coloured light mixer

C. Smith, Manchester College of Education

There seems to be a wide range of colour mirrors on the market, which are often costly and of varying effectiveness. The cheapest and most effective which I have used was designed and constructed by our laboratory technician, Mr B. Turner.

The essential requirements are a piece of wide thick cardboard tubing with a tin lid to fit, although with a little more skill and patience a coffee tin could be used instead. The lid simply has a lampholder bolted to its underside so as to take a 12 V lamp (Figure 1). The tube has three holes of 3 cm diameter bored near its base, as

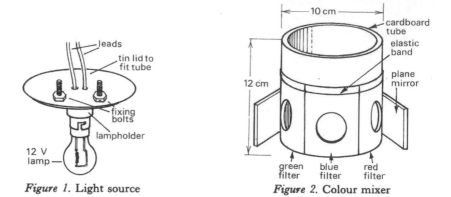

Figure 1. Light source

Figure 2. Colour mixer

shown in Figure 2. Each of the holes has a piece of coloured filter covering it, held in position by a rubber band. When the apparatus is assembled and standing on a bench, with two angled plane mirrors positioned as shown, coloured areas are thrown on to a screen in front, and these can easily be overlapped by adjusting the mirrors.

Although I have used it only for small group demonstration purposes, the mixer is safe, cheap and robust enough for pupils to use to investigate the overlapping of a range of colours, colour shadows and the appearance of suitable small coloured objects in lights of different colours.

Physics for Robert Burns

F. R. Dean, W. R. Tuson College, Preston

Even physically, we do not 'see ourselves as others see us'. In the mirror our image is, in the ordinary way, laterally inverted: if one has a hair parting on the left, the image has it on its right. Even our features may look different to us compared to how other people see us, for how many faces are truly symmetrical? Are eyebrows always the same level on both sides? Is one's smile always equally divided? Wrinkles? Cheek bones? One could go on. If you stand with your back to the television and view through a mirror, do familiar faces look quite so familiar? And let us just note in passing, that in the mirror we can look ourselves squarely in the eyes, despite the inversion.

In physics we learn about two mirrors inclined to each other. We see multiple images. If the angle $\theta°$ between the mirrors divides exactly into 360° we see $360/\theta - 1$ images.

Figure 1. One mirror

Figure 2. Two mirrors

Section 3: Optics

Go up to a mirror on the wall and place a hand mirror at an angle to it slightly less than 90°. Four images are seen. The two outer images, each seen after a single reflection, are laterally inverted. The inner two, each seen after two reflections are not laterally inverted. These two inner images cannot look you in the eye. Now gradually increase the angle. These inner images merge into one—one which *does* look you in the eyes. We now have, of course, exactly three images. The angle is exactly 90°. Checking, $360/\theta - 1 = 3$. Correct.

We are now able to see ourselves, physically at least, as Robert Burns wished. The fused central image is not laterally inverted. Give a wry smile on the left, and the image smiles back wrily on the left—its left, that is, and not our left, which is what we are accustomed to seeing.

Tunnelling light waves

John Marshall, Christ College, Brecon

At the start of Chapter 10 of his book *Atomic Physics* [1], Born gives an introduction to 'The size of the nucleus and α-decay'. An analogy is drawn between the escape of an α-particle from the potential crater of a nucleus, and the breakdown of total internal reflection at a glass/air interface which occurs when a *second* glass surface is brought close to it. A simple way to demonstrate this is as follows. Two good quality 60° glass prisms are cleaned and then put face to face. If the air is excluded, by wringing the two surfaces together, a substantial patch will now transmit white light in the direction shown in Figure 1.

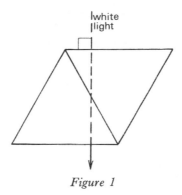

Figure 1

When the prisms are very slowly and gently eased apart, the intensity of the transmitted light falls quickly but not abruptly and the light becomes redder, rather like the setting sun. Red light, then, can tunnel its way through a wider air-gap than blue light.

REFERENCE

1. Born, M., *Atomic Physics*, 8th edition (Blackie, 1969).

A worksheet on light for upper primary/lower middle school pupils

Rennie Macleod Wilson, Alness Academy, Ross-shire

INTRODUCTION

The worksheet formed the first part of a three-part topic with a general theme 'colours'. The topic was taught to P7 pupils in the five feeder primaries of Alness Academy, namely, Ardross, Bridgend, Coulhill, Kiltearn and Obsdale Primary Schools. The choice of subject was determined by four main factors:

1. Avoidance of overlap with S1/S2 science at the Academy and the inclusion of elements of the three secondary sciences.
2. Convenient division into practical teaching units of approximately an hour, as eight classes were to be fitted into the available free periods of the secondary teacher (over two terms).
3. Avoidance of any extraordinary expense by the primary schools and of exceptional facilities not usually found in primary schools (although equipment might be borrowed from the Academy).
4. Independence of weather conditions (i.e. an indoor topic) as these can be unpredictable—snowfalls in May, for example.

METHOD

Sources of light, e.g. sunlight, light bulb, were elicited. Pupils described their colour as white/yellow. A projector and prism were used to show the colours of light when split up and the parallel with the rainbow was drawn.

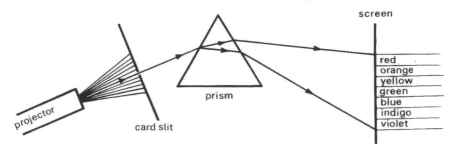

Figure 1. Light split up into seven colours (spectrum)

Hypothesis was formed that light only appears white/yellow to us because the colours are blended in some way. Pupils then made a Newton's disc and spun it. The colours changed from the seven rainbow colours to a single colour (off-white). Various suggestions were made to explain this, e.g. the eye saw each colour so quickly that it could not separate them, hence a single blended colour.

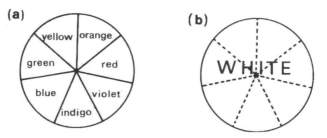

Figure 2. Newton's disc (a) still, (b) spun

Pupils then looked at light through filters—red, green and blue, noting the effect. Next, pupils looked at light through overlapping filters and noted the colours as turquoise (blue/green), yellow-brown (red/green) and violet (blue/red). Overlapping of all three filters usually produced black (although differences in filter strengths occasionally gave other colours—see note at end of article.

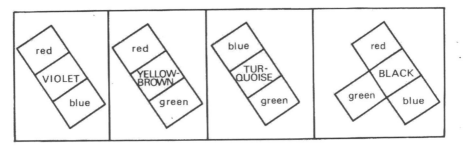

Figure 3. Colours of light seen through overlapping filters

Pupils then drew four squares, the first of which contained a blue flower with green leaves in a red pot. They drew the colours they could see when looking at this picture through red, green and blue filters, noting that the filter eliminated its own colour each time.

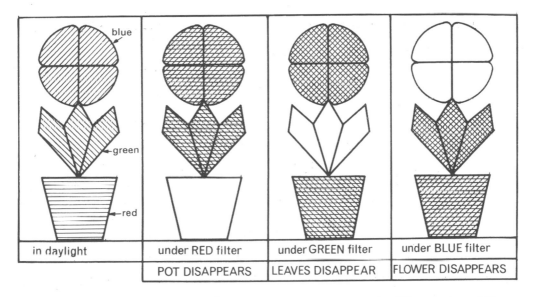

Figure 4. Picture of red pot, with blue flower and green leaves, seen under different coloured filters

The pupils then used two discs, using both sides, to see if the colours they had seen with overlapping filters could be made by spinning discs. Each side was quartered and coloured red/green, red/blue, green/blue and finally, red/green/blue. These discs were spun and the colours were seen to match the overlapping filter colours, except the red/green/blue, which was white, not black. (It was suggested that this difference might be caused by the different filter strengths, since the three-filter overlap had not invariably given black, but had sometimes shown one or other of the filter colours to be just distinguishable.)

LETTER COLOUR →	RED	ORANGE	BLUE	PINK	BLACK	BROWN	PINK	GREEN	YELLOW	BLUE
'MYSTERY WORD'	C	A	W	R	A	L	L	E	N	S
under GREEN filter	C	A	W	R	A	L	L	/	(/)	S
under BLUE filter	C	A	/	R	A	L	L	E	N	/
under RED filter	/	/	W	/	A	L	/	E	/	S

Figure 5. Coloured quartered discs (red/green/blue) (a) still, (b) spun

Pupils finally used their filters to solve 'mystery' cards which had been made for them. These cards were lists of places, people, everyday items etc., and the name could be found by trying the filters in turn over the coloured letters. Some letters would be eliminated each time, and one filter would eliminate the letters in such a way that the remaining visible letters spelt the 'mystery' word. Although lists of nouns were used, messages can be made and have been successfully tried also.

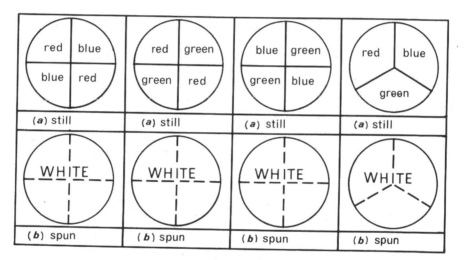

Figure 6. Specimen of 'mystery' word under different filters

CONCLUSION

The description and the diagrams in this worksheet required two–three lessons to complete, as far as the practical component was concerned. With regard to the writing up of the work, this was left to the primary teacher to arrange as desired. The worksheets themselves, which appear here in condensed form, could be filled in by the pupils, but it was also possible for the pupils to write up the experimental work and draw the diagrams for themselves and in the format preferred by their teacher. It was also possible for the teacher to develop the themes further if he/she wished and/or to integrate them into a particular, individual scheme of work. The primary teachers helped in the teaching of the practical work and both they and I gained a valuable insight into each other's teaching methods.

NOTE

The filters used correspond to Griffin & George XGB–440, colours 010L (Red), 030F (Green), 050W (Blue). The colours used in the discs and for the 'mystery' cards are made by colouring pencils—12 assorted, purchased from John Menzies. It may be noted that the paler the colours are drawn, in general, the better the effects, e.g. with all discs.

The inverse square law for light as an analogy with that for gamma radiation

D. P. Newton, Spennymoor Comprehensive School, Durham

Most will be familiar with the standard textbook demonstration of the inverse-square law for gamma radiation in which the radioactive source is placed at various distances from the detector (usually a ratemeter). The same experiment can be done with a small solar cell [1] and, for example, a 200 W, clear light bulb, as shown in Figure 1.

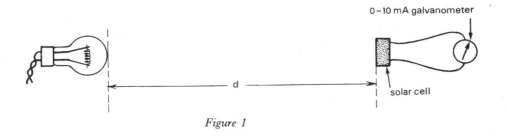

Figure 1

Following the procedure for the gamma radiation experiment, the background 'count' is first taken with the bulb off. This is subtracted from subsequent galvanometer readings. Distance d is measured and the reading taken with the lamp on. d is varied and current readings I are recorded, corrected for background light. A graph of $1/\sqrt{I}$ against d completes the analogy to demonstrate the inverse square law. Figure 2 shows what can be obtained by an unassisted sixth former. Note that, as with the geiger-tube, there is a negative intercept on the d-axis.

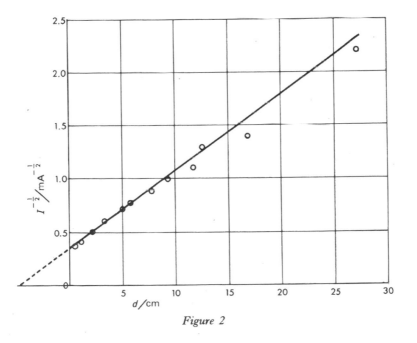

Figure 2

It should be added that the experiment assumes that the cell's response to light, within the limits of this experiment, is a linear one.

REFERENCE

1. RS Components Ltd.

Additive colour mixing

Randal L. Henly, Mount Temple School, Dublin

I was interested to see R. A. Firth's account of his method of demonstrating colour mixing [1] which to me is quite novel. He also mentioned using three slide projectors containing respectively red, green and blue filters; this is certainly an ideal method, and if each projector is fitted with a dimmer (stage light dimmers are ideal), most instructive and interesting demonstrations can be carried out, far beyond the requirements of school physics.

However, assembling three projectors at one time inevitably requires arrangements and the co-operation of colleagues. One of my teaching policies is to carry out as many demonstrations as possible which require minimum preparation (we don't have laboratory technicians in this country), and to make as much use as possible of equipment which is always at the ready, such as the overhead projector. The following method of demonstrating colour mixing may therefore be of interest to readers.

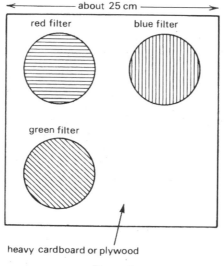

Figure 1

A three-colour filter is prepared (Figure 1). This is placed on the stage of the overhead projector, so that three separate coloured beams can be projected. The projector is turned around so that the beams are projected away from the screen. Three plane mirrors are attached to a retort stand, using clamps and bossheads, and the whole assembly is placed so that the mirrors catch the three coloured beams from the projector and reflect them back to the screen Figure 2. It takes a little bit of manipulation to set the mirrors so that the reflected beams clear the lens assembly of the o.h.p. and appear on the screen as complete circles of light.

Now by swinging and/or tilting the mirrors slightly, any two colours can be made to overlap (giving the various secondary colours), all three can be mixed to give white, or all combinations can be shown together, by arranging the overlapping pattern as shown in Figure 3.

Figure 2

The same apparatus can be used to show the appearance of coloured objects in light of various colours. The projector is turned back to its usual position, and two of the coloured filters covered over. By holding various coloured objects in the light beam, it can be shown for example that a red object viewed in green light appears black (or at least, very dark). Incidentally, it is a good idea to spend a bit of time collecting coloured objects (or books etc.) which 'work well' for this demonstration, and to label them or keep them solely for this purpose. A darkened room is necessary for both this and the previous demonstration.

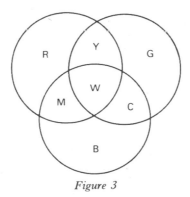

Figure 3

There are so many varied, interesting and instructive demonstrations, which can be done with the overhead projector, that it would make a useful article if anyone were prepared to write it.

Section 3: Optics

A 'magic' bubble motor

Alan Ward, College of Saint Paul and Saint Mary, Cheltenham

The picture shows an effective bubble motor that I have just made up. It is mounted on a 'pillar' made from a Domestos bottle with its top cut off, on top of which is a saucer, to support an inverted milk bottle. (The 'dip' in the milk bottle's bottom is convenient.) Supporting arms for the motor are formed from about 60 cm of stiff iron wire. The 'motor' itself is a yogurt pot with a hole (to receive a pea-shooter) bored approximately 3 cm from the opening. Just before use, it is hung, upside down, by attaching it to a short roll of BluTak fixed to an end of the wire. I have used a little padlock as a counterweight, hooked to the other end of the wire.

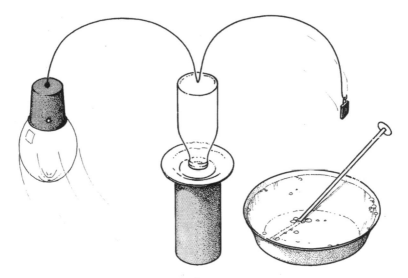

To prepare the motor, attach the pea-shooter, dip the inverted pot in a strong solution of liquid detergent, and blow a large dangling bubble. Carefully pull out the pea-shooter, keep a thumb over the hole, and hang the pot-and-bubble on the wire—making sure that the hole (which I made with a chemist's cork-borer) is pointing 'along a tangent' to the potential circle of rotation of the device. Gently let go, taking care to dampen any vibration of the wire—and wait a moment. Ever so slowly at first, the motor starts to drive itself—by jet-propulsion—and it gathers speed, as the bubble gets smaller (and pressure of the air inside increases) [1]. It should completely circumnavigate the milk bottle.

REFERENCE

1. Ward, A., 'An unusual parallelogram of forces', *S.S.R.*, 1976, 203, **58**, 335.

A hot-water bottle and a (solar) 'serpent'

Alan Ward, College of Saint Paul and Saint Mary, Cheltenham

A safe way to experiment with heat is to use a hot-water bottle. It can be brought into a primary school classroom, wrapped up in a blanket—and discussion can begin at once, about using water for storing energy, and on the value of wool as an insulator. One use for the hot-water bottle is to wrap around a glass bottle that has a balloon stretched over its neck, to show that air expands when it is heated. My most amusing use so far is to provide an up-draught of hot air, to operate a gaily decorated, stiff-paper spiral toy 'serpent' that I suspend from the ceiling, via cotton and BluTak. The model stops rotating when I hold a tray over the heat source, suggesting that it is something coming up from the vicinity of the hot-water bottle that is making the serpent spin. When air in the room is cold, I find that the paper snake rotates slowly above my bared warm arms. So I can get it to work with body heat.

Discussing the serpent with teachers, I was challenged to prove that it was *air* ascending above the hot-water bottle. First I thought that it would be appropriate to demonstrate that an upward breeze would turn the spiral. Using a battery-driven fan, I showed that blowing air upward made the serpent turn clockwise (and then fanning air downwards made it go anticlockwise). Then we sought a way to 'mark' the air with a tracer; we tried banging a dusty chalk-eraser with a stick. Motes of dust by the side of the serpent moved about in random ways, although it was possible to observe definite upward swirling of the chalk motes in the air column above the hot-water bottle. A spotlight might have made this effect more easily visible. The teachers agreed that our 'playing about' was in the tradition of good primary science.

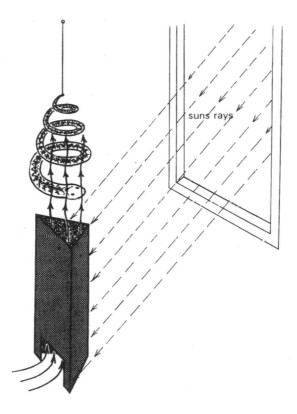

suns rays

A simple solar camera for measuring the sun's diameter

Alan Ward, College of Saint Paul and Saint Mary, Cheltenham

At noon, sometime last June, when rays from the sun were least affected by atmospheric refraction (light bending from a straight path during its passage through successive air layers of increasing optical density), I used a black paper square with a 'pinhole' through it, a metre rule, centimetre-squared graph paper and some blobs of BluTak 'putty', to get a rough measurement of the sun's diameter.

15 × 15 cm squares of black paper and graph paper were used; the hole in the middle of the black paper being made with the point of compasses. The squares were fixed at right angles, either end of the ruler, using the BluTak. So I could assume that the pinhole and calibrated 'screen' were 100 cms apart. I had constructed a crude pinhole camera.

Pointing the black card at the sun, through the window, around noon, I observed the image of the solar disc projected on the shadow cast over the square-divided screen—and I was able to estimate that the rather dim circle was about 0.9 cm in diameter. (In summer time woods on sunny days, 'holes' in the leaf cover produce the effects of a 'multiple-pinhole camera', projecting clusters of solar images on woodland tracks.)

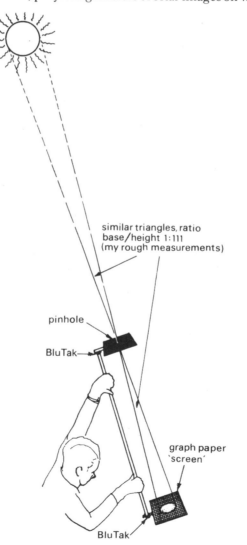

similar triangles, ratio base/height 1:111 (my rough measurements)

pinhole

BluTak

graph paper 'screen'

BluTak

I assumed that the diameter of the sun 93 000 000 miles away formed the base line of a triangle with its apex at the pinhole. Also I knew that the other side of the hole was the apex of a similar triangle (i.e. similar in proportion) having a 0.9 cm base line one metre away. Dividing 100 cm by 0.9 cm gave me the ratio 1:111. Dividing the (approximate) distance to the sun by 111 should give me a fair idea of the solar diameter.

93 000 000 miles is about 150 000 000 kilometres (and one decimal place too many for my pocket calculator to handle). Dividing 93 000 000 by 111, I obtained a number of 837 837 miles for the sun's diameter. Its actual diameter is 865 370. I was out by a mere 27 533 miles . . . But actually this was an error of only about three per cent. Of course I made my calculation from the sun's average distance from the earth, and I was rather lucky. But my method gave me the feel of how it was possible over hundreds of years, for astronomers to make celestial measurements, using very simple apparatus.

Mathematically able older juniors should be able to understand the experiment, after experiences with homemade pinhole cameras. Use the device described to observe a solar eclipse *indirectly* (the apparatus can be vastly improved). Millions of 'indirect' solar eclipses are observable in summer woods . . . *Warn children never to look directly at the sun.* By the way: 865 370 miles is 1 384 592 kilometres.

Demonstration of no parallax and image formed in a plane mirror

John Forden

No parallax is always a difficult concept to explain, and consequently many students, including sixth formers, find it difficult to understand. The demonstration described has been found to be a simple, but very effective way of overcoming this.

With a lamp (eg 12 V 24 W) about a metre in front of a perspex safety screen, a reflection can clearly be seen in a moderately lit laboratory.

The position and properties of the image seen on the screen can then be discussed.

To find the exact position of the image the idea of no parallax can be introduced. As the screen is transparent the background can be seen clearly through it, with a little imagination it appears as though a duplicate, but laterally inverted, lamp is behind the screen. An object such as a laboratory stand can then be used to 'search' for the image. The teacher moves the stand, on the opposite side of the screen to the lamp, until the pupils, from any part of the room, can see the stand in exactly the same position as the reflection of the lamp. This is the position of no parallax.

We have found that this is a very good way of introducing an experiment that the pupils then can carry out themselves, to determine the position and properties of an image formed in a plane mirror. This is achieved using pins in place of the lamp and stand described above.

However, the deeper understanding of no parallax gained has wider benefits, and is useful during the study of optics as a whole.

J Forden, Roch Valley High School, Cornfield Street, Milnrow, Rochdale, Lancs.

Rutherford's red box

Stephen Rutherford, Pimlico School

Use this box in a circus of experiments which reveal the nature of light. If only red light is present, only 'red' can be seen.

> Look through the hole in this box and press the switch. Now tell me the colours of the Lego bricks. Now change the order of the bricks and ask your partner to guess!

All you need is a shoe box. Fix a PP3 battery to the lid with a sticky pad and use it to light a red led (light emitting diode). The light is near enough monochromatic, and the experiment shows very clearly that the colours we see depend upon the wavelengths present in the light as well as in the surface properties of the material.

Perhaps the secret is in the led. Use the sort sold by R. S. Components as 'ultrabright', a pack of three red ones in clear plastic casings costs about £2. They give much more light than the 'usual' led's. Sadly only red led's are suitable.

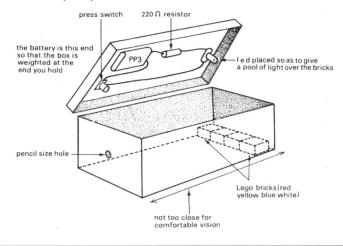

press switch   220 Ω resistor

the battery is this end so that the box is weighted at the end you hold

PP3

l e d placed so as to give a pool of light over the bricks

pencil size hole

Lego bricks (red yellow blue white)

not too close for comfortable vision

A demonstration of the production of colour on a TV screen

J Davy

Any science course which deals with the topic of colour is likely to include a description of how a TV screen is covered with thousands of dots or bars which glow red, blue or green when struck by beams of electrons, and how the colours which appear on the screen are built up from a combination of these three primary colours. The dots or bars are too small to observe easily with the naked eye, but if a microscope of fairly low power (about × 30 - a travelling microscope is ideal) is mounted horizontally in front of the screen they can be seen clearly.

The program for a BBC computer which follows enables the colour of the screen to be changed by pressing the numbered keys 1 to 8. The appearance of the colour on the screen can

then be readily compared with the combination of dots or bars which are glowing.

```
10   MODE 1
20   ON ERROR RUN
30   *KEY 10 OLD I M RUN I M
40   VDU 28,0,12,39,2
50   VDU 24,100;100;1179;500;
60   VDU 19,1,0;0;
70   GCOL 0,129:CLG
80   VDU 23,1,0;0;0;0;
90   PRINT "PRESS NUMBER FOR COLOUR
        REQUIRED"
100  PRINT ""1 BLACK      4 YELLOW
        7 CYAN"
110  PRINT "'2 RED        5 BLUE
        8 WHITE"
120  PRINT "'3 GREEN      6 MAGENTA"
130  REPEAT
140  G=GET: IF G<49 OR G>56 THEN 140
150  X%=G-49
160  VDU 19,1,X%;0;
170  UNTIL FALSE
```

J Davy, Siddal Moor High School, Heywood, Lancs

Fibre optics - a low cost method

AR Sparkes

Fibre optics kits (such as Maplins) and fully constructed units are available from various scientific suppliers. They all have two drawbacks, cost for a class set is high and they confuse the pupil with apparently complex electronics. I needed very low cost and simplicity. I also wanted to demonstrate the effect of cladding and bend radius.

The fibre optic shown (Figure 1) is a length of 10 mm diameter acrylic rod (Perspex), bent

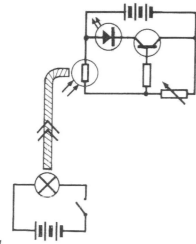

Figure 1

through a right angle. This can be used with either the Locktronics equipment shown in Figure 2, or with the MFA logic boards many of us now have. In order to use the LDR on the MFA it is necessary to couple it to the light guide. This was done with a small tube the same diameter as the LDR, a 2 cm wide strip of black paper wound round a clamp stand bar and fastened with tape formed a satisfactory tube. The input end of the rod can be held close to the light source by hand or anchored with a small piece of plasticine. In conditions of high ambient lighting, it may be necessary to enclose the bulb and the end of the rod in plasticine in order to exclude stray light input.

The rod, price £2.62 per 2 m length (thinner diameter at lower cost could be used but would be less robust), was purchased from: VT Plastics, 14 Midland Street, Manchester M12 6LB (061 273 1732). It was processed as follows:

Cut to length. I use 50 cm lengths. Polish the end faces. If you cannot get the craft department to do this for you, the following procedure will get good results. Using first a coarse grade emery cloth or similar, and then a fine grade, placed flat on the bench; square and smooth the faces of the vertically held rod. Place a piece of absorbent cloth on to a flat surface and impregnate it with a liquid metal polish. Polish the ends of the vertically held rod on this cloth.

The rod should then be bent. Again, if you do not have access to the right equipment, it can be done in a saucepan. Bring a saucepan full of water to the boil. Hold the end of a rod in the boiling water. A slow steady pressure to bend the end over will eventually succeed. Do not hurry this. You are only just above the softening point of the acrylic and it has poor thermal conductivity hence it takes several minutes for the heat to penetrate the rod fully. Once you have achieved the bend that you want, continue to boil for about half an hour in order to remove the stress from the bend. If this is not done, fractures will occur later. Quench in cold water to set the shape. This produces a rod with a typical radius of curvature of about 3 cm. This is way below the normal recommended minimum radius of curvature of twenty times the fibre diameter, but above the minimum for the axial ray of two times the fibre diameter (see JMB Physics Syllabus A, Advanced Paper 1, Section 2 question C2 for 1988).

The following experiments are performed with this light guide:

1 Either (A): Allow the pupils to experiment with the rod on its own. They should discover that if one end is pointed towards a bright light, the other end is seen to be bright. This should lead to the idea of light travelling along the rod round the bend.
or (B) Pupils have a container of water (a plastic lunch box) to which they add a few drops of milk to make it just cloudy. The light

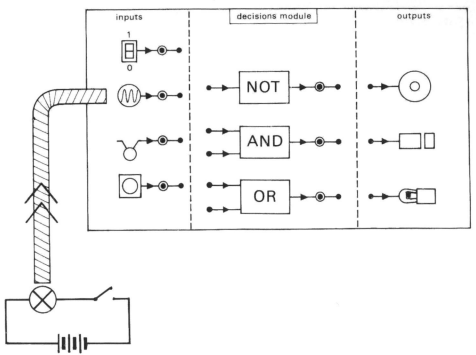

Figure 1

guide is then set up with a strong light close to one end, eg a bare torch bulb, and the other end dipped into the cloudy water. With low ambient lighting the light can be seen coming out of the end of the rod and producing a diffuse beam in the 'milk'.

2 Pupils set up the apparatus below to produce a Morse transmission system. With Locktronics alone, you may have to be satisfied with the receiver as a flashing LED. With MFA the light sensor can also be connected to the buzzer. Because of the low radius bend, it is necessary to clad the rod on the bend to prevent the ingress of light from the surroundings. This must be done in such a way that there is a layer of air between the rod and the surroundings. If adhesive tape is used, the refractive index of the adhesive is sufficiently high to prevent total internal reflection. The cladding can be done using approximately 1 m length of 10 mm wide plastic strip cut from a black dust-bin liner or some other similarly opaque, narrow material wound onto the rod. The need for this cladding can be demonstrated, at the end of the experiment, by displacing some of the wraps on the bend and noting that the detector is permanently on; ie light is entering the

rod other than at the end. A length of tight fitting PVC or rubber tubing pushed onto the rod does not work. After a while, the PVC or rubber seems to make intimate contact with the material of the rod and the transmission drops off considerably.

3 Demonstration only. I have a signal generator with an output able to supply a low voltage bulb with variable square wave. A CRO across the receiver LDR produces a fairly good trace.

4 For more advanced students, the need for cladding, the effect of the low radius bend, immersing the bend in white spirit/turpentine, etc can be investigated in more detail.

AR Sparkes, Macclesfield County High School, Macclesfield.

Understanding parallax

Geoff Auty

After many years of teaching parallax (in order to understand how to look for no-parallax), by asking the class to look at two misaligned rulers, I hit upon a more interesting approach using a plastic measuring cylinder which has been left on the bench from a previous lesson.

As shown in the diagram, a ruler was clamped well away from the pole of the clampstand, with the measuring cylinder beneath but just behind the ruler as viewed by the class.

As I set this up whilst discussing something else, the class took little notice of how the apparatus was assembled.

I asked the remainder of the class to look at one student near the centre of the room who happened to be sitting in line with the ruler and cylinder. I then asked that student to say whether the ruler would fall into the cylinder when the clamp was released.

Not only did he move his head considerably from side to side before passing judgement, but naturally, the other students looked towards the ruler to see what he was expected to assess, and to see if they could do better, so that quite soon and without prompting, the students in the centre of the room were all rocking in their seats as they looked at the apparatus. Students at the sides of the room did not need to do so. However, before releasing the ruler, I asked those students whether the measuring cylinder was to the right or to the left. Those on the left said it was to the left (as in the Figure 1). Those on the right had a different view.

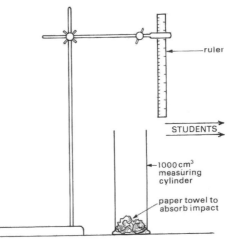

ruler

STUDENTS

1000 cm³ measuring cylinder

paper towel to absorb impact

Figure 1

I released the ruler, predictably it missed.

Then I clamped it again but directly over the cylinder. Now all the class were rocking in their seats and realizing that the lines of the ruler and cylinder never separated.

Now they understood parallax, they could describe it as the apparent movement of two fixed objects due to lateral movement of the observer; with the further object seeming to move the same way as the observer. More important, they were also ready to attempt with confidence those optics experiments which require accurate judgement of no-parallax for image location.

Geoff Auty, New College, Pontefract, West Yorkshire

A cheap light source for ray streaks

I Lawrence

As part of a refurbishment of the class experiment kit for years 7 to 11 we have developed a cheap and reliable ray streaks light source. These can be made for £3-4 each and are good for all of the Revised Nuffield O Level Physics year three experiments up to and including two different coloured sources through a model astronomical telescope (experiment 31a).

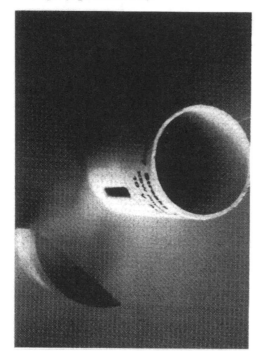

The construction is from an 80 mm length of 40 mm diameter white waste pipe. The slit to allow the light out is drilled/filled or routed out and three holes drilled to centre the festoon lamp holder opposite the slit in commercially available clip lamp holders. The festoon lamp should be mounted as close to one end of the tube as possible to allow for the maximum 'throw' along the benchtop. Two further small holes allow thin flexible wires to supply the electricity.

We use 12 V festoon lamps (available from car accessory shops) but have found that they produce adequate luminous power and last much

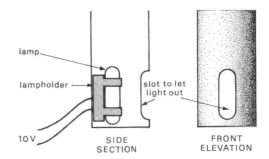

longer if run at 10 V.

The thin wire permits the low mass sources to be accurately positioned. The tops of the tubes have been left open to allow the pupils enough ambient light in a darkened laboratory to read instructions!

I Lawrence, The King's School, Worcester

'Sound-on-light'—a transmitter and receiver

J. R. Smith and J. Willis (Science Sixth Form), Eggar's Grammar School, Alton, Hants

It was decided to experiment with the transmission of sound by using a light source which would enable communication over a short distance, thereby demonstrating the properties of a phototransistor.

The light source used was a torch bulb, rated at 6 to 8 V and 6 W. The sound source was a tape recorder with internal speaker disconnected and the bulb in series with a 6-volt battery joined across the extension loudspeaker sockets (Fig. 1).

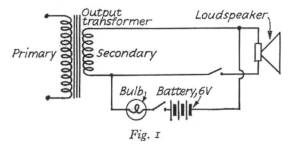

Fig. 1

The tape recorder was then switched on, the bulb glowing brighter or dimmer in accordance with the strength of the signal. The bulb was placed in a tube, at the focal point of a concave mirror ($f = 10$ cm) so that the emitted light was concentrated into a parallel beam (Fig. 2). A light-sensitive transistor was used as the receiver (in our case an OC 71 with the paint removed, but since these are now manufactured with an opaque case, an OCP 71 may be used.)

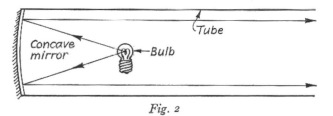

Fig. 2

The transmitted light was directed on to the transistor by a convex lens. It was later found that the most sensitive part of the transistor was the tip (Fig. 3).

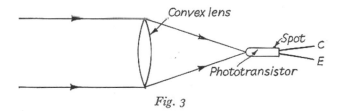

Fig. 3

The output of the OC 71 was connected to a single-transistor pre-amplifier and then into the radio-tuner circuit of an amplifier (Fig. 4). Screened cable should be used for connecting photo-transistor to pre-amplifier, and pre-amplifier to main amplifier.

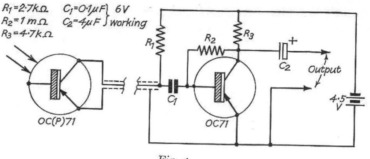

Fig. 4

The output of the amplifier was fed into a loudspeaker and the transmitted signal heard. Background 'hiss' was introduced on the pre-amplifier and a 100 cycle hum was heard if a mains light was switched on.

It was found that the unit worked well over a distance of 20 yards but great care was needed to align the source and receiver. The reproduction over such a distance was reasonably clear, but background noise was becoming disturbing.

The apparatus worked successfully in daylight, but it made alignment increasingly difficult.

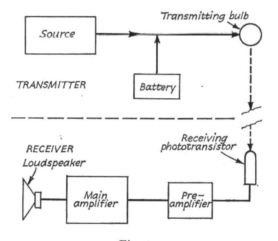

Fig. 5

A block diagram of the complete arrangement is shown in Fig. 5. Fig. 6 shows the transistor connections.

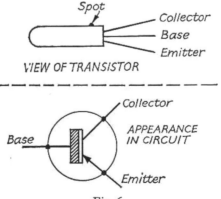

Fig. 6

Resonance and beats with a ticker timer

J. M. S. Whittell, Nairobi School, Kenya

A standard ticker timer (e.g. Philip Harris) can be used very simply to show both resonance and beats.

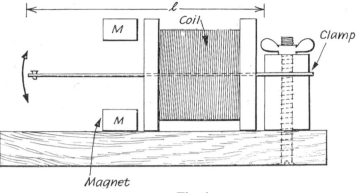

Fig. 1

The anvil is removed so that the end of the vibrator blade can oscillate freely. With the coil connected to the normal 6–12-volt a.c. supply, the double amplitude of the vibrator is measured with a millimetre scale as a function of the total vibrating length l (see Fig. 1).

The results are plotted in Fig. 2a and show an obvious maximum at a vibrating

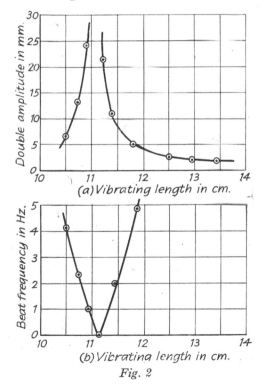

(a) Vibrating length in cm.

(b) Vibrating length in cm.

Fig. 2

length of about 11 cm. Near to this length the amplitude of the oscillations is limited by the internal diameter of the coil.

In addition, if at each setting the end of the vibrator is tapped with a finger, the resulting vibration shows a beat. This is due to the superposition of the forcing frequency (50 Hz) and the natural frequency of the vibrator. Measurements of the beat frequency are necessarily pretty rough, particularly as the resonant length is approached. However, some results which have been obtained are plotted in Fig. 2b, and do at least show that the beat frequency tends to zero as the amplitude approaches a maximum.

Demonstration of beats using mechanical and audio vibrations

K. W. Sharratt, Chelmsford Technical High School

The beat effect of two mechanical vibrations as well as that of two audio oscillations may be demonstrated using loudspeakers. The speakers used are 5 in (12.7 cm) diameter, 3 Ω speakers, one supplied from the mains via a 6 V transformer at 50 Hz, the other from a signal generator.

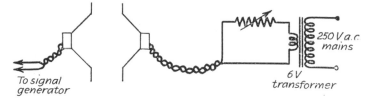

With the frequencies nearly matched the speakers are held facing each other about $\frac{1}{2}$ m apart; beats are difficult to discern. When the distance is reduced, audible beats become apparent. When the speakers are held in physical contact, the beats are well pronounced and can be seen clearly as mechanical vibrations of the speaker diaphragms. The frequency of these may be as low as one every 3 seconds with an amplitude of up to $\frac{1}{2}$ cm.

Change of beat frequency can be demonstrated by adjustment of the signal generator; the beats are so well heard and seen that with the aid of a large clock a class can do quantitative work on the relationship between the speaker frequencies and the beat frequency.

Standing waves on a metre rule

W. H. Jarvis, Rannoch School, Perthshire

We have found that good standing waves can be demonstrated by mounting a metre rule at its 50 cm mark to a vibrator (ours is supplied by Messrs Linstead). We used a Philip Harris audio oscillator, which can deliver more than 5 W into 3 ohms.

At a nominal 16.5 Hz, the resonating rule showed two antinodes, and a node, surprisingly, at the centre point. At a nominal 60 Hz, there were antinodes at each end and in the middle; and at a nominal 137 Hz, there were five antinodes including one at each end.

Assuming some calibration error, my first-year sixth have learnt a lot in attempting to explain the harmonic relation between these three modes of vibration.

Many other nodes are rendered visible, if light powder is sprinkled along the rule.

The double beam oscilloscope: sound and waves in elementary physics
W. K. Mace, King Edward VII School, Sheffield

INTRODUCTION TO THE OSCILLOSCOPE

Children work in pairs, each pair having a pencil (or better, a felt pen) and a long strip of paper such as could be provided from a roll of decorator's lining paper sawn in half. One child executes 'vibrations' by moving the pen rapidly to and fro across the paper, whilst the other draws the paper steadily along the bench. By asking them to vibrate rapidly, slowly, widely, jerkily, etc., they can observe how the 'shape in time' of a vibration can be made visible as a shape in space; regularity, frequency, amplitude and 'waveform' can be identified in terms of their characteristics in the trace.

The next step is to connect both a vibrator unit and the CRO to the same AFO, using the lowest available frequency and a very slow time-base: they can see their 'pen traces' being precisely drawn across the screen. By increasing frequency and sweep speed until persistence of vision takes over, they can be shown exactly what the oscilloscope is doing.

Having added to the above an extra emphasis on the fact that 'right means later', a simple and accurate way of measuring the speed of sound becomes possible.

THE SPEED OF SOUND

As illustrated in Figure 1, microphones A and B, connected respectively to Y_2 and Y_1, pick up a 10 kHz signal from a speaker. A is kept stationary and the oscilloscope is set to trigger on A. If B is now moved away from the source, the Y_1 trace moves to

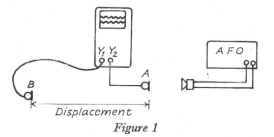

Figure 1

the right as the vibrations arrive increasingly late at B. By counting the number of peaks passing a fixed point on the screen one can tell how many ten-thousandths of a second late the signal is. At about 35 cm it is ten vibrations late, which amounts to $\frac{1}{1000}$ s, giving a speed of 350 m/s.

SUPERPOSITION OF WAVES

This is frankly a *simulation* demonstration, and this must be firmly emphasized. It is intended as a reinforcement or clarification after everything possible has been shown with real waves.

The CRO is externally triggered, conveniently using the 2 V output from a mains transformer unit. If then a signal at 49 Hz is applied to Y the trace will move slowly forwards at the rate of one 'wave' per second. 51 Hz will produce a wave moving backwards. The same effects are obtained using, for example, frequencies each side of 200 Hz. Given two AFOs (one of which must not be earthed), Y_1 and Y_2 can display any two moving waves, either indistinguishable in wavelength, or in a simple ratio to one another. Further, by arranging the switching system shown in Figure 2 the sum of these waves can be displayed on one trace; thus one can study in simulation the spatial relationship of two moving wave trains and the result of their superposition.

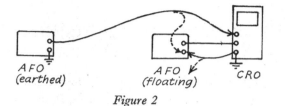

Figure 2

A good thing to show first, to make it clear what the device is doing, is a high frequency of small amplitude superposed on a large amplitude low frequency, the waves moving in the same, and also in the opposite directions. Figure 3 shows the result of using $50 - 1$ Hz and $450 + 9$ Hz.

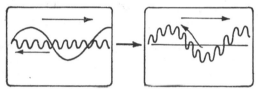

Figure 3

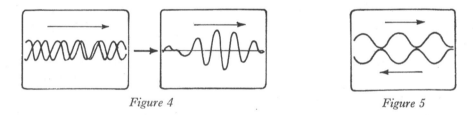

Figure 4   Figure 5

Beats can be shown using, for example, frequencies close to 250 and 300 Hz; these can be held steady or made to move slowly across the screen, first separately, then superposed (Figure 4). The formation of standing waves can be looked at very fruitfully by causing two waves of (sensibly) equal wavelength to move across the screen very slowly in opposite directions either on the same line or one immediately above the other (Figure 5). By this means it will be seen that crests coincide only at certain fixed points in space, and the resultant slow standing wave is quite beautiful to watch.

For more advanced pupils it is instructive to look at the effect of unequal amplitudes on beat and standing wave patterns, especially the latter, which is a curious blend of standing and progressive wave.

Sonometer experiments

R. Sen and R. McGarva, first year sixth-form, Mander College, Bedford

Being two unmusical first-year A-level students we experienced difficulty with the classical sonometer experiments (using a tuning fork and sonometer). Our difficulties were increased by the second harmonic obtained when the tuning fork was held on the sonometer sounding board. Whilst trying to resolve this situation we devised the following experiment.

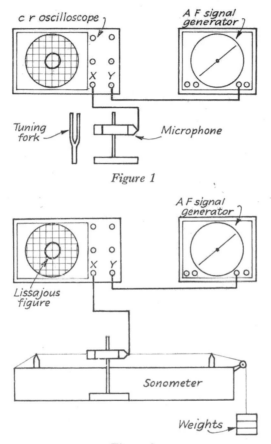

Figure 1

Figure 2

A tuning fork was held near a microphone and its note was fed into the X input of the oscilloscope. At the same time a signal from an AF signal generator was fed into the Y input (Figure 1). By adjustment of the AF signal generator frequency, a near circular Lissajous figure was obtained. This indicated matched frequencies and is shown in Figure 1. In the second stage (Figure 2) of the experiment the sonometer replaced the tuning fork and its signal was fed into the microphone and then to the X input of the oscilloscope. By adjustment of the length of the wire under test, a Lissajous figure similar to the first Lissajous figure was obtained. This indicated that the frequency of the wire was the same as the AF signal generator, which in turn was the same as the tuning fork. The length of the vibrating wire was then recorded. This was repeated using tuning forks of various frequencies.

In our particular experiment, to show that $F \propto 1/l$ eight points were plotted giving a very good straight line passing through the origin. Note that the AF signal generator was not a replacement for the tuning fork.

Another easy way to show slow 'beats'

W. H. Jarvis, Rannoch School, Perthshire

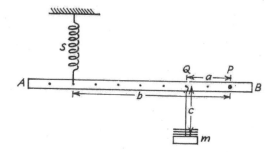

AB is a metre rule, fixed at P by a pivot, and hanging from a Nuffield-type spring S. A variable mass m hangs at a distance a from P.

Two types of oscillation are possible simultaneously:

1. the mass m can swing as a compound pendulum pivoted at Q;
2. the ruler and mass can execute oscillations in a vertical plane about the pivot P.

It is easy to adjust these two modes of oscillation to have nearly the same frequency by varying a, b, c or m. In this condition, the transfer of energy at the difference frequency between the two 'modes' can easily be seen.

I am grateful to sixth-former Niall Taylor for bringing this effect to my attention.

Young's experiment without effort

A. G. Reed, Dr Williams' School, Dolgelley, Merioneth

All too often I have found that setting up a demonstration of Young's experiment has been time consuming and not very convincing. I was surprised when I realized that I had never seen any reference to the use of a spectrometer for this purpose.

The collimator need only be roughly adjusted, together with a sodium source, to give good illumination of a suitable pair of slits mounted in place of the table. The slits may now easily be aligned at right angles to the incident light. The objective lens of the telescope is removed and the eyepiece adjusted. A good number of fringes can now be seen after suitable alignment and adjustment of slits.

The double slit may readily be turned through a known angle to demonstrate the effect of decreasing the apparent slit separation. The diffraction caused by an adjustable slit, straight edge and pin may also be shown with the minimum of effort.

Commercially available double slits, adjustable slit and pin can be obtained mounted on '¼ inch' rod, allowing direct replacement of the spectrometer table.

Simulating wave motion by a flickerbook

C. Smith, Manchester College of Education

An understanding of waves requires the motion to be slowed down sufficiently to see both the overall motion of the wave itself and of the individual particles of it. Few middle schools possess either the commercial models for demonstrating wave motion or have facilities for loop films. The flickerbook technique effectively replaces these models, whilst at the same time providing each pupil with his own model, which he has made himself and which has no distracting hidden mechanisms. Thus a worthwhile exercise for a group (or indeed a whole class) is to make flickerbook simulations of wave motion.

Ordinary exercise books can be used for this especially if they are turned on their side to use the printed lines as reference lines for drawing in the curves and dots. To be effective at least 20 different diagrams have to be drawn and the movement must be small between each page. This means the diagrams have to be quite large (2–3 cm high). The drawing of the diagrams to the required standard is not easy for pupils with little experience of wave motion, and consequently they require considerable guidance. This can be provided by utilizing workcards from which pupils *trace* the diagram by placing the card under the page and moving it along a little for subsequent pages. The following workcard explains the process.

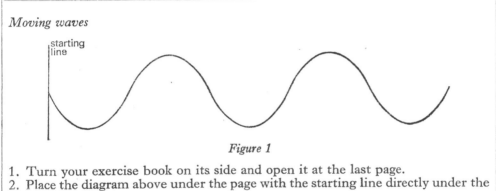

Moving waves

starting line

Figure 1

1. Turn your exercise book on its side and open it at the last page.
2. Place the diagram above under the page with the starting line directly under the bottom line.
3. Trace the pattern.
4. Repeat on the next to last page with the starting line one-third of a line along, but only trace up to the same line as before.
5. Move the diagram so as to trace the 'missing' third.
6. Do the same on the second to last page starting two-thirds of a line in.
7. Continue for at least 20 pages.

Once the drawing is completed, one particular crest can be crayoned in a different colour on subsequent pages and its motion observed. Heavy dots can then be placed at the intersection of the curve with a particular line throughout, thus showing the up and down motion of the particles. The whole process is easier to do than to explain, so once pupils have done a few pages there is in fact little difficulty.

As well as showing the motion, the completed book can be used to see the relationship between velocity and frequency for constant wavelength. A 'slow' flicking shows a low frequency and hence a small velocity, and vice-versa for a quicker flicking. If this relationship is to be pursued further at this stage a separate wave train can be drawn on the same pages, but having a different wavelength for the same starting point.

The same technique can be used for longitudinal waves, the book being turned over and a wave train drawn on the opposite pages.

Demonstration of beats using mechanical and audio vibrations

K. W. Sharratt, Chelmsford Technical High School

The beat effect of two mechanical vibrations as well as that of two audio oscillations may be demonstrated using loudspeakers. The speakers used are 5 in (12.7 cm) diameter, 3 Ω speakers, one supplied from the mains via a 6 V transformer at 50 Hz, the other from a signal generator.

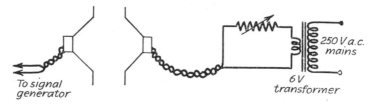

With the frequencies nearly matched the speakers are held facing each other about $\frac{1}{2}$ m apart; beats are difficult to discern. When the distance is reduced, audible beats become apparent. When the speakers are held in physical contact, the beats are well pronounced and can be seen clearly as mechanical vibrations of the speaker diaphragms. The frequency of these may be as low as one every 3 seconds with an amplitude of up to $\frac{1}{2}$ cm.

Change of beat frequency can be demonstrated by adjustment of the signal generator; the beats are so well heard and seen that with the aid of a large clock a class can do quantitative work on the relationship between the speaker frequencies and the beat frequency.

Scanning diffraction patterns

W. K. Mace, King Edward VII School, Sheffield

A difficulty for some A-level students is that of identifying a given diffraction pattern as seen on a screen with the corresponding intensity profile as plotted on a graph. I had toyed with the idea of putting a photocell in the diffraction pattern from a laser but abandoned it, until at a Sheffield University Physics Department Open Day I saw it done, using a rotating mirror (why didn't I think of it!) and a storage oscilloscope with a light-operated trigger system. The result was outstandingly worthwhile, but did demand the use of more complex equipment than is to be found in schools. The arrangement described below shows how the same thing can be done with standard school apparatus.

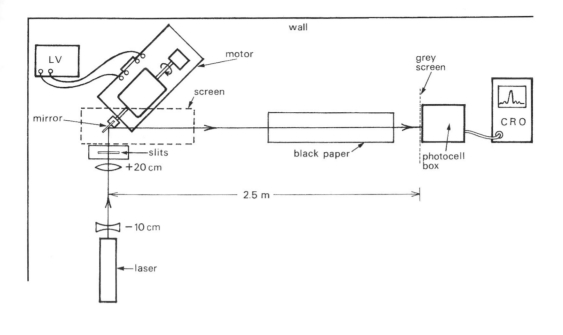

Figure 1

Figure 1 shows the general layout, to which the following comments refer.

1. The motor has to have a shaft free from play or vibration, otherwise the beam is liable to miss the photodiode. The fractional h.p. motor is adequate in this respect.
2. The optimum rate of turning seems to be about 5 revolutions/s (standard watch-tick). The motor is series connected and run off an l.v. power unit.
3. The mirror has to be good quality—but nowadays most are. It need not be front-silvered. I use an ordinary 2.5 cm × 7.5 cm O-level class mirror.
4. The mirror fixture could be Araldite, but I find the plastic plug from a Griffin & George thermometer case can be used perfectly well by clearing the inside with a sharp knife and cutting a groove in the end with a hacksaw.
5. The scan is vertical, so diffracting slits need to be horizontal. This makes position adjustment of the slits a matter of raising or lowering, and a Philip Harris holder is perfect for this purpose.
6. The beam rotates about the motor axis and therefore cannot flash across the spectators. However, DES specifies the use of screening and this must of course be fitted.
7. All benches are shiny at grazing incidence, hence the black paper to eliminate spurious reflection signals. An alternative would be a suitably placed black 'screen with hole' from the Physical Optics kit.
8. The purpose of the lenses is to provide a beam wide enough to accommodate several slits, and to focus it on the photodiode.
9. The screen at the photodiode enables the diffraction pattern to be looked at as a whole, complementary to the scanning demonstration.

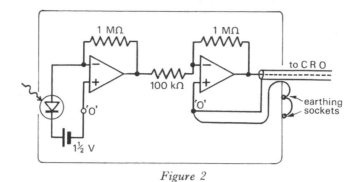

Figure 2

Figure 2 shows the display circuitry. The photodiode is RS Components No. 305–462. Its sensitive area is less than 1 mm square, so it can resolve fine detail well. When reverse biased its conductivity is proportional to light intensity. The operational amplifiers are from the Unilab electronics kit; the first gives a negative output proportional to intensity, whilst the second inverts and multiplies by 10. The oscilloscope is the normal school type, but satisfactory display depends crucially upon good triggering.

Where I work the diode picks up radio stations, and I find it necessary to put the whole assembly inside a box. If a half-size biscuit tin is available there is plenty of room for everything including two PP9s to power the amplifiers.

The trace repeats five times a second, and with 'brightness' to maximum this gives a very acceptable picture in the dim-out conditions suitable for looking at diffraction patterns.

Single slit diffraction may at first sight be disappointing, but adjustment of gain and timebase speed will enable the subsidiary maxima to be seen. One has to remember that intensity is proportional to (amplitude)², and that the intensity of the first subsidiary then works out as about only 1/20 of that of the central maximum. The second is about 1/60, but with care can just be seen.

Young's fringes show the diffraction 'envelope', and one can see them as far as the first subsidiary maximum by turning up the gain until the central fringe is off the screen. Three slits show the single subsidiary between pairs of maxima, and four show two subsidiaries. At five I find the pattern a little distorted and could not convincingly claim three subsidiaries. This may be a difficulty due to trying to pack five slits into an inadequate beam width: I certainly find in all patterns that a minute upward or downward movement of the slits can effect a considerable improvement.

This is a most satisfying demonstration. On pure theory one works out the energy distribution in detail and draws graphs of rather beautiful shape. To take a real source of light and see these graphs perfectly reproduced is quite an exciting experience.

Speed of sound in a steel rod

W. K. Mace, King Edward VII School, Sheffield

A steel rod clamped loosely at its mid point rings clearly when tapped at its end, giving the fundamental mode longitudinal standing wave of wavelength equal to twice the length of the rod. If this frequency can be measured it provides an accurate measure of the speed of sound in the rod and hence of the Young's modulus. Nuffield A Experiment 4.7 notes this and suggests measuring the frequency by tuning (by ear) a loudspeaker to the same note. For the sake of the tone deaf it would be much preferable to excite the vibration directly from a calibrated AFO, but because of coupling problems this proves to be almost impossible if one tries, for example, to use an ordinary vibrator unit. It can be done, however, by magnetically polarizing one end of the bar and using a surrounding coil to excite the vibrations. A convenient arrangement is shown in the diagram.

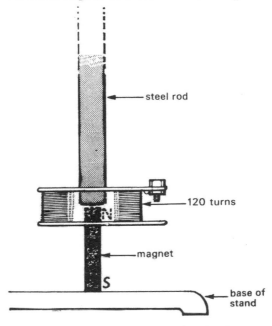

The rod is held at its mid point in a padded clamp so that its lower end is about a millimetre away from the magnet. The coil is separately clamped and fed from the power output of the AFO. The resonance is loud and can be pinpointed to within less than 1 Hz, which of course is much smaller than the AFO scale can resolve. If one wishes to take this opportunity of exploiting a potentially very high precision measurement it is very easy to measure the frequency using a scaler. A squarer and two triggered bistables from the electronics kit divide the frequency by four and the result is accepted by the coax socket at the back of the Panax scaler. Only three minutes of counting then gives the frequency to one part in 1000, which is comparable with the uncertainty of measuring the length of the rod with a metre rule. Pupils may further find it interesting to reflect that by counting for two days against an ordinary electronic stopwatch one could measure the frequency to one part in a million—if it stayed that steady!

The use of a guitar in vibrating-string experiments

Bill Harvey, Napier College, Edinburgh

There are several contexts in which vibrating strings or wires appear in school physics. Normally a sonometer is used. I would like to argue the case in favour of the guitar as a supplement, or indeed an alternative, to the sonometer.

ADVANTAGES OF THE GUITAR

Unlike the sonometer, the guitar is a genuine musical instrument, and, moreover, one which many pupils may readily associate with their own musical tastes. If one of our aims in teaching physics is to relate school studies to 'real life', it seems a good idea to demonstrate the physical principles involved in something as commonplace as a guitar as well as in the unfamiliar, artificial (and unmusical) sonometer.

There are also some technical advantages in using a guitar. It produces clear, pleasant tones which die away less quickly than those of a sonometer. Guitars provide a set of easily adjustable parameters (mass, length and tension of strings) which would require either several sonometers or restringing. Nylon string guitars, in which the lower strings are wirewound, give a particularly vivid demonstration of the effect of changing mass on the

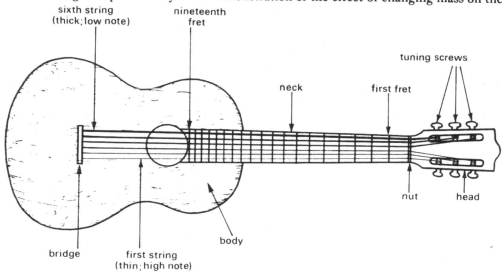

note produced. (A possible disadvantage of the guitar is that a quantitative value for tension is not easily obtained.) Length is not only easily variable (by pressing a finger behind one of the frets) but because of the built-in frets, measurements of length are easily made. The connection between these preset lengths and the frequency ratios between musical notes can be discussed if desired.

EXPERIMENTS

1. Factors affecting the frequency of a vibrating string

The role of mass, length and tension are easily demonstrated. Frequencies can be checked by standard methods. In particular, halving the length produces a note one octave higher (pressing the string at the twelfth fret) and this can be shown for six different frequencies without retuning. If the open string is plucked, then shortened by bringing a finger sharply down behind a fret (the 'hammer-on'), the note is not damped; it simply changes to a note of higher frequency.

2. Harmonics

It is easy to show that plucking a string at different points gives rise to notes of varying tonal quality. (Mellow tones result from plucking at the neck, and tinny, sharp tones from plucking near the bridge.) Also, the guitar always sounds much less 'flat' than a signal generator driving a loudspeaker at the 'same' frequency. The role of harmonics can then be discussed.

Several harmonics can be generated by plucking a string while *lightly* touching the string at certain specific frets (i.e. by creating a node), then removing the finger. The string then vibrates along its whole length yet clearly emits a higher than normal note. For example, a node at the twelfth fret produces a note one octave higher than the fundamental. (The same note, produced for a different reason, is obtained by pressing firmly behind the twelfth fret as described above.) Other harmonics can be produced, for example at the fifth, seventh and nineteenth frets. These effects work best with the heavier strings.

3. Resonance

Resonance is often demonstrated using a tuning fork and a tube containing a variable height of water. The guitar provides a further illustration which, in my experience, is more selective (a smaller frequency band produces resonance) and provides more vivid resonances. For example, by knocking on the body of the guitar the strings can be made to vibrate at their fundamental frequencies. A further demonstration of the selectivity of the strings can be performed by holding a tuning fork on the bridge of the guitar and altering the tension, or the effective length, of the strings. Resonance can be heard, and can be seen if a V-shaped piece of paper is placed on the string; the paper jumps and buzzes.

Another method of generating a driving frequency is to use the next lowest string, and progressively shorten it. (The guitar need not be correctly tuned, but one advantage of so doing is that one has a clear idea of what will happen!) For a correctly-tuned guitar, the fourth string will resonate when the fifth string is plucked while held down at the fifth fret. Nothing at all happens at adjacent frets. To emphasize the point, the fourth string can also be driven by the sixth string held down at the tenth fret.

It is also possible to show resonance at frequencies other than the fundamental. For example, one can show that the sixth string, held down at the nineteenth fret, sounds the same as the open second string. Next, one plucks the second string with the sixth string left open. If the second string is then damped, the sound persists. By a process of elimination (damping each string in turn) this sound can be traced to the sixth string, resonating at a frequency more than double its fundamental. A similar relationship exists between the first and fifth strings. And so on.

CONCLUSIONS

I have tried to argue that the guitar provides a useful method of investigating the properties of vibrating systems. Other applications no doubt exist, and the reader is invited to explore the possibilities. Quite reasonable guitars can be bought for under twenty pounds, and tutors dealing with tuning, stringing, etc., can be had for under a pound. Alternatively, enquiries would probably reveal a surprisingly large number of guitar owners among colleagues and pupils, some of whom may be prepared to assist.

A simple model ear drum

J. R. Nicholson, formerly Paedagogical Institute, Nicosia

The bottom of a plastic cup is removed, perhaps with a hot scalpel, in such a way as to leave a smooth edge. Cling film is stretched across the opening, and secured to the sides of the pot by smoothing it down firmly. The film resonates when a child sings into the pot, rather like the membrane between the outer and middle ear. The resonance can be felt with the finger tips as shown in Figure 1. When the film is broken by pushing a pencil through it, the film no longer vibrates—unless the hole is very small. This might be a salutory warning to those who push pencils and other things into their ears!

The resonance of the membrane can be demonstrated visually as shown in Figure 2. A

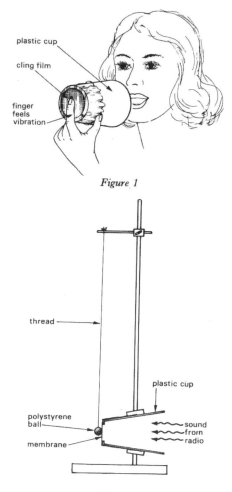

Figure 1

Figure 2

small polystyrene ball, suspended by a thread, rests against the centre of the film. It is kept in position by gravity, the thread being almost, but not quite, vertical. To avoid arguments about wind from singing throats and vibrations through the laboratory bench the sound can be provided by a hand held radio. The ball responds to the vibrations by bobbing about in a random manner.

Speed of sound in carbon dioxide

C. B. Spurgin, formerly Wolverhampton Grammar School

The recent note by P. Caines and D. F. Ford under the above heading (*S.S.R.*, 1982, 225, **63**, 730–2) reminds me of a much simpler version which I used to use when velocity of sound in gases, and the factors affecting it, featured more strongly in syllabuses than they

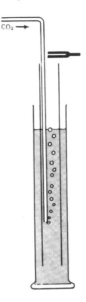

do now. Using the tuning-fork and resonance-tube method, it was sufficient to pass CO_2 from a generator—we used a CO_2 Kipp's apparatus—via a long delivery-tube below the water surface in the resonance-tube. The denser CO_2 displaced the air and was kept flowing until constancy of resonance-length showed that the tube was full. The CO_2 was no drier than air usually is in such an experiment, and the end-correction region may not have been full of pure CO_2; but satisfying values were found very quickly, both for speed of sound in the gas and for the ratio of its principal heat capacities.

The wave projector

Michael J. Williams, Barstable School, Basildon, Essex

A projection of a thick copper spiral is produced on a screen at about 1 to 2 m distance from a small-filament car bulb. There is magnification and distortion but the moving wave can be seen as the spiral is turned by a low-volt motor.

If the screen is replaced by a strip of card then the vertical oscillation of any point is seen. Two cards at a distance from each other show phase difference. Measurement of time and distance can 'confirm' the equation linking frequency, velocity and wavelength.

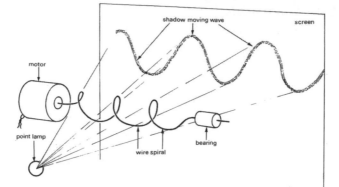

The lamp must, be shaded, of course, from the class to avoid glare. Varying distances will vary wavelength and amplitude. The spiral is wound tight onto a glass tube of radius 3 cm and then carefully extended evenly and its ends turned to form centre axes. To link to the motor I use an electrical strip connector, the other end turns in a plastic cotton reel.

A cheap and simple vibration generator

J. Oxford, Audley Park School, Torquay

With the cost of proprietary vibration generators running at around fifty pounds, it is well worthwhile considering a cheap and simply constructed homemade apparatus.

The main component is a moving coil speaker of 4 Ω impedence and of about 20 cm diameter or 20 cm along the major axis if an eliptical speaker is used. To the centre of the cone is cemented a light weight pillar of aluminium, plastic or wood using epoxy resin glue and of such a length that it stands above the rim of the speaker cone by about 2 cm. The outer end of the pillar is drilled, tapped and slotted to accept the vibrating cord and a small clamping screw.

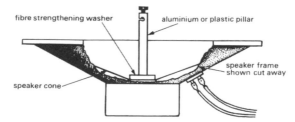

Figure 1. Detail of modified speaker

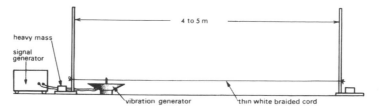

Figure 2. The vibration generator set up to demonstrate the standing wave experiment

If necessary the speaker can be mounted in a simple frame although many speakers are stable and heavy enough to remain free standing. The vibration generator will operate in the vertical or horizontal position and is controlled by a standard laboratory signal generator. Care is needed when in use to ensure that no undue side strain is placed upon the vibrating pillar.

Use a syringe for an instant vacuum

Michael Kahn, University of Botswana

A neat way of showing that sound does not pass (easily) through a vacuum is provided by using a syringe. The technique described here works for both glass and plastic syringes. The best vacuum results with a large glass syringe, but a 25 cm³ plastic one is capable of giving a vacuum down to 30 kPa if the plunger is rapidly withdrawn while the nozzle is closed with a finger.

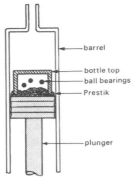

The only item now needed is some kind of rattle which can be enclosed in the vacuum. This can be improvised by attaching a small plastic bottle top to the rubber piston by means of 'Prestik'. Three small ball bearings are placed inside this top to act as a rattle. Naturally the top must be small enough to enter the barrel, and this clearance and the damping of the Prestik and piston ensures little sound is transmitted to the outside directly through the walls of the syringe by transmission (see figure).

Evacuate the syringe. Shake well. Little sound is heard. Remove finger. Whoosh. Shake. Lots of sound is heard.

CD reflection grating

Michael Brimicombe

A compact disc makes a good reflection grating. Light of wavelength λ which hits the grating perpendicular to its surface can leave it with an angle of reflection θ if $n\lambda = d\sin\theta$ (see figure). My sixth form students have estimated the spacing between the grooves d with the following experiment.

1 Place the compact disc on the bench. Arrange a lamp so that its filament is about 10 cm above the grooved surface of the disc, as shown in the figure. A ray box lamp (12 V, 24 W) will do. Clamp a 50 cm rule so that it is vertical with its base 15 cm away from the portion of the disc immediately under the lamp. You will need to mark that point on the disc with a felt tip pen.

2 Look at the surface of the disc so that the light passes by the edge of the rule. Start off at the 50 cm mark and move down until you find the first order ($n = 1$) spectrum.

3 Measure the value of x for yellow light. You can assume that yellow light has a wavelength of 590 nm. Use the value of x to calculate a value for the groove spacing d.

4 Make a similar measurement for the second order ($n = 2$) spectrum.

We obtained a value of about 1.6 micrometres.

M Brimicombe, Cedars Upper School, Leighton Buzzard, Beds

Investigating the properties of infra-red radiation using a television and its remote control

Peter Hooker

When studying the properties of electro-magnetic radiation most teachers confine themselves to 'light'. However many households have an emitter and detector of infra-red radiation, which pupils are familiar with, so why not use them? Pupils can be asked to investigate the various properties for themselves, a real Sc1 that can be done for homework! or you can go through the investigations below as a class experiment.

The television should be set to Teletext and the sound 'muted' or turned down, then every time a number key is pressed the television will give a 'beep' showing that it has detected the infra-red message sent by the remote control. We now have an emitter and a detector of infra-red radiation.

POSSIBLE INVESTIGATIONS

1 Does the infra-red radiation from the remote control pass through a sheet of paper? If so how many sheets?
(The answer is yes, in my experiments, about 3 sheets)

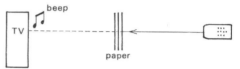

2 We know that light passes through glass, does infra-red radiation?

3 Does infra-red radiation pass through clear plastic?

4 Does infra-red like light travel in straight lines?

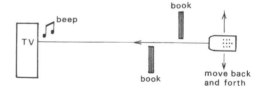

5 Can infra-red radiation be reflected?

6 Do the normal laws of reflection apply?

7 Can we make an infra-red periscope?

SUMMARY

Having carried out our experiments with a remote control we can now compare some of the properties of two different electro magnetic radiations, infra-red and light.

Property/Radiation	Light	Infra-red
absorption	will pass through glass	will pass through glass
absorption	will pass through clear plastic	will pass through clear plastic
transmission	travels in straight lines	travels in straight lines
reflection	angle of incidence = angle of reflection	angle of incidence = angle of reflection

P Hooker, Kent Curriculum Services Agency

Instant magnetic fields

G. R. Williamson and I. Hilton, Wellacre County Secondary School, Flixton, Urmston, Manchester

Magnetic fields made with iron filings can be rendered sufficiently permanent for classroom display by spraying with hair lacquer. This is considerably quicker and simpler than the traditional waxed-paper method.

A simple demonstration of a circuit breaker

Alan Ward, St Mary's College, Cheltenham

Electromagnetic circuit breakers are used instead of fuse wires in power stations and factories to 'switch off' electricity supplies when short circuits occur, or when circuits are 'overloaded' with current. The following easy-to-make model demonstrates a circuit breaker in action.

Begin by making a solenoid: wind 100 turns of 28 s.w.g. insulated copper wire, in layers, to cover an inch near one end of a short piece of thin glass tubing. Insert drawing-pins A and B, an inch apart, in a scrap wood base-board. Sellotape the solenoid half an inch in front of one of the pins. Join the solenoid to the nearest pin and to a terminal of a $4\frac{1}{2}$-volt bell battery. Join the second pin to a bulb-holder H (containing a $3\frac{1}{2}$-volt torch bulb) which is screwed to the board. Then join the other side of the bulb-holder to the free terminal of the battery. Open one prong of a large steel paper-clip, and poke the prong into the solenoid, whilst resting most of the clip across pins A and B. The circuit is completed and the light goes on.

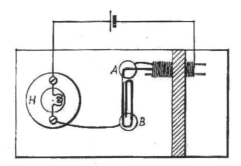

To operate the circuit breaker, hold the bared ends of a spare wire across the screws on the bulb-holder. The light goes out, thus reducing the resistance in the circuit and letting a 'large' current flow around the solenoid. When this happens, the solenoid's force increases and pulls the paper-clip off the pins. The circuit is broken and the electricity turned off.

Hertzian waves to make a neon bulb flash

J. C. Siddons, Thornton School, Bradford

In 1888, less than 80 years ago, Hertz fitted brass plates to the terminals of the secondary of an induction coil. When the induction coil was set working, Hertz noticed sparks, though only very small ones, across the gap in a loop of wire held near. Each induction coil spark had set in motion electromagnetic waves which, after travelling through space, produced a second spark. 'Wireless' had begun.

The second spark is very faint: it requires great patience to produce it. Hertz's historic experiment is not at all suitable for classroom demonstration. The Hertzian waves released by the spark from a Van de Graaff generator will produce a bright flash in a nearby neon bulb to which an appropriate voltage is already applied. This modernized version of Hertz's experiment serves well as a classroom demonstration.

The 'beehive' type of neon bulb, being large, is suitable. As is well known, a neon bulb has a 'sparking voltage' below which it will not light up but above which it lights up brightly. Once the neon bulb has lit up, it remains lit even when the voltage is reduced below the sparking voltage. With continued reduction in voltage the lamp eventually goes out when the 'extinction voltage' is reached. It is a simple matter to find the sparking voltage of a particular neon bulb, not quite so simple to find its extinction voltage.

Apply a voltage lower than the sparking voltage but higher than the extinction voltage to a neon bulb placed near to a Van de Graaff generator. With the latter, produce a single spark. The neon bulb will light up and will remain lit. The pulse of waves released by the spark momentarily pushes the voltage of the neon above the sparking voltage.

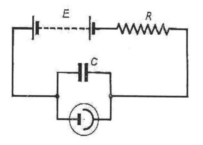

It is a more interesting experiment to watch if the neon bulb is made to flash instead of lighting permanently. This can be done by using the simple CR arrangement shown in the figure. Suitable values are C, 1 microfarad; R, 200,000 ohms. The neon bulb is now controlled by the voltage across the condenser. The condenser is charged by the battery to a voltage E which is made to be below the sparking voltage. Now when the Hertzian waves from the Van de Graaff spark cause the neon bulb to light, the condenser partially discharges. Its voltage drops below the extinction voltage of the neon bulb and so the light goes out. The Van de Graaff spark has produced a neon flash. The battery now charges the condenser up again through the resistor R and so after a short interval of time the neon bulb is ready to respond to the next spark. (Sixth-formers can use the experiment to study the time constant, CR.)

Attempts to use the neon flashes to show the reflection and polarization of the Hertzian waves have so far not been successful.

'Like' and 'opposite' electric charges

Alan Ward, St Mary's College, Cheltenham

Stretch the top of a nylon stocking, before pinning it, still stretched, towards one end of a 1-ft stick. Repeat, using a second stocking and stick, to complete a pair of the devices. Also, tie an inflated 'round' toy balloon on each end of a 4-ft length of wool.

Hold a stick with its limp stocking dangling, and stroke the nylon from top to toe and on both sides, using one of the balloons. Doing this charges the nylon with static electricity, which causes it to fill out dramatically as if containing a phantom leg.

Let an assistant hold the stick attached to the charged stocking. He must stand in an open space and hold the apparatus well clear of himself, otherwise the stocking will be attracted towards him.

Charge the second stocking (as described), using the second balloon. Give the second stocking-on-a-stick to the assistant and ask him to hold both sticks side by side. At the same time, hold up the middle of the wool connecting the balloons.

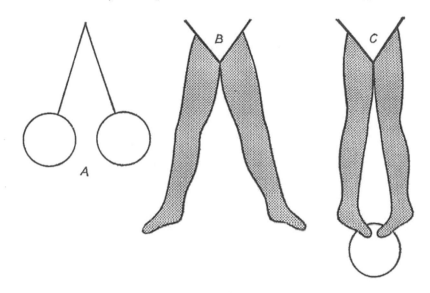

The stockings (B) repel each other—so they must carry 'like' charges. Also, the balloons (A) repel each other—so they too must carry like charges of static electricity. But are both sets of charges alike?

If one of the charged balloons is brought up between the feet of the charged stockings, the feet are attracted (C). Indeed, by moving the balloon up and down rhythmically, the stockings can be made to 'dance'. Since attraction occurs between opposite (i.e. negative and positive) electric charges, the charges on the balloons must be opposite to the charges on the stockings.

Rubbing a plastic lunch-box with a woollen scarf charges the box negatively. It will be found that such a negatively charged box repels a freely-suspended charged balloon. It follows that the balloons carry negative charges—and that the stockings carry positive charges.

Demonstration of a magnetic field pattern in three dimensions

G. Auty, The King's School, Pontefract

This has been achieved using an 'Eclipse Major' horseshoe magnet, with 500 g iron filings sprinkled on to a glass sheet (the front from a balance case) placed just above it.

When very few iron filings are sprinkled on to the glass, they settle along the smooth curves of lines of force, tracing the directions from north to south, along the glass sheet. Straight lines are evident between the poles; the filings directly above the poles stand up vertically.

As more filings are sprinkled on, they form into a heap above the magnet, and show

Fig. 1. Oblique view of field pattern in three dimensions

the curves tracing lines of force above the glass in addition to those along it. A very large field pattern in three dimensions can be built in this way.

Some lines are complete curves, but many are not complete, extending only to the limit of the iron, and finish in mid-air. All two-dimensional field patterns in textbooks show such incomplete lines. By very carefully adding more filings to the centre of the pile, it is possible to show that lines can be completed which were previously incomplete.

When incomplete lines reach the end of the iron available, they pass into the air, and join the iron again near the south pole. The air is of course not noticeably affected by the magnetic field passing through it.

Fig 2. Plan view showing the lines of force in all directions, particularly those standing up near the poles

If the magnet is turned through 90 degrees, the bulk of the iron filings does not turn with it. The filings in view reorientate themselves to form the pattern in its new position. This illustrates that the magnetic field can be moved through the iron without moving the iron bodily.

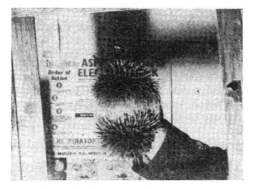

Fig. 3. The glass sheet is sandwiched vertically between the magnet and the filings; the demonstrator is holding the magnet only

One can lay further emphasis on the strength of this effect, and on the fact that the position of the field depends on that of the magnet, by holding the magnet only. It can be turned so that the glass sheet is vertical and the iron filings occupy the same position relative to the magnet as they did when the sheet was horizontal.

Improvement to the 'water circuit'

W. H. Jarvis, Rannoch School, Perthshire

In the usual water analogue of an electrical circuit, a 'voltmeter' is represented by two manometer tubes across the 'resistance' or pump (as Figure 1).

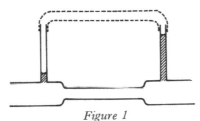

Figure 1

The problem is that pupils are asked to accept that there is nowhere in the electrical circuit where electrons can enter or leave; yet here we have two openings through which water can, and sometimes does, spurt. This is answered, and the similarity to a voltmeter at the same time improved, by adding a clear plastic pipe shown dotted in Figure 1.

Teaching the potentiometer

M. L. Cooper, Thurrock Technical College, Grays, Essex

The null-point condition of balance for a potentiometer quite often causes bother for students; the following approach may be of interest. It tackles the problem via both practical and theoretical considerations, and assumes that the students appreciate the concept of electromotive force.

1. A voltmeter is used to investigate the potential drop along a potentiometer wire, the teacher having first ensured that the total drop along the wire is capable of measurement with the available meters (Fig. 1). This will supply the linear relationship of voltage drop proportional to length of wire.

2. The voltmeter is replaced by a cell, a suitably protected galvanometer, and sliding

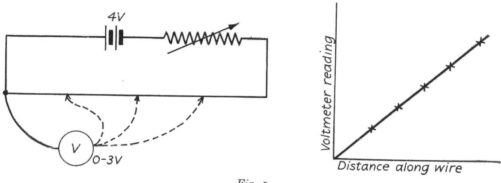

Fig. 1

jockey. The position on the wire giving no deflection is found, having first checked that the deflections at the two ends of the wire are in opposite directions. From the balance length and the graph from the first experiment, the e.m.f. of the cell is found.

3. A further calculation would be to include an ammeter in the potentiometer circuit, and to use the current flowing and the resistance/cm of the wire (found from a separate experiment) to calculate the voltage drop/cm for the wire. This value can be compared with that obtained from the slope of the graph.

Since, at balance, the galvanometer reading is zero, no current is being taken from the cell, so the p.d. across its terminals must be its e.m.f. To verify mathematically that this p.d. is the same as the voltage drop across the balance length of wire, two calculations are performed.

4. Apply Kirchhoff's second law to this hypothetical circuit (Fig. 2), i.e.

$$E_1 = (x + y)R + xr_1$$
$$E_2 = (x + y)R + yg + yr_2.$$

These are completely general; now consider the special case when the component values can be chosen to make $y = 0$, i.e. so that the galvanometer reading is zero.
Then

$$E_1 = xR + xr_1$$
$$E_2 = xR.$$

But the voltage drop across R is xR; thus, when $y = 0$, the e.m.f. E_2 is balanced by the voltage drop across the resistance R.

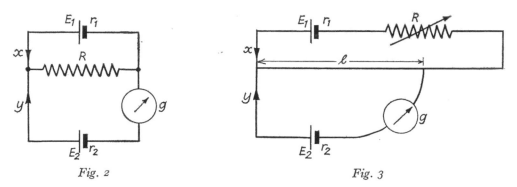

Fig. 2   Fig. 3

5. Perform the same sort of analysis for a general potentiometer circuit (Fig. 3), with the wire having a resistance of r ohm/cm.

i.e.
$$E_1 = (x + y)lr + x(100 - l)r + xR + xr_1$$
$$E_2 = (x + y)lr + yg + yr_2.$$

When balanced, i.e. $y = 0$,
$$E_1 = xlr + x(100 - l)r + xr + xr_1$$
$$E_2 = xlr.$$

Thus, the e.m.f. E_2 is balanced by the voltage drop (xlr) across l cm of wire.

Although this approach may be of use, there is, of course, no substitute for an understanding of the concepts of potential at a point, potential difference, and electromotive force, if students are to appreciate this basic piece of electrical equipment.

Induction coil

T. C. Johnston, Wilson's Hospital and Preston School

By connecting together in series a 4.5 V battery, an electric bell, and the low voltage winding of a mains bell transformer, a simple and obvious induction coil is produced, the output being at the mains terminals of the transformer. Besides the simplicity and low cost of the arrangement it has the advantage that the pupil sees the contact breaker and transformer as the separate basic components of an induction coil, and can therefore more easily understand the functioning of the large demonstration model, if such there be.

Phase tester bulbs in electrostatics

T. C. Johnston, Wilson's Hospital and Preston School

The small neon bulbs used in phase testers give pupils individual experience of the fact that static and current electricity are but different forms of the same phenomenon. Charged balloons and bits of plastic, and of course an electrophorus plate, will all cause them to light up. Besides this, since it is the negative electrode that lights up, they form a useful test for conventional sign of charge. In more advanced work they may be used to show the presence of an electric field and equipotential 'surfaces' around an electrically charged conductor.

A teaching aid that will help to explain the electric motor

Alan Ward, Saint Mary's College, Cheltenham

For a pupil to understand how a simple d.c. electric motor works, it is important for him to realize that, during every half-rotation of the armature, the current in the armature changes direction. With less able children the point might need some emphasis. Here is an easily improvised device that can be used by teachers to help clarify the idea.

On thin white paper, draw a diagram representing the armature and commutator (*a*). Put in arrowheads to show the direction of the current—'as we look down on the machine'. Reverse the paper and redraw the diagram, as if superimposed, on the opposite side of the paper (*b*). Again indicate the way the current is going by putting in arrowheads.

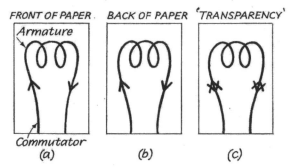

The diagrams represent how the current must flow in the armature on the two occasions when it is horizontal during a complete rotation. To the casual eye nothing seems to have happened to the current. But, if the paper is held up against bright light and viewed as a transparency (*c*), it will be obvious that the current in the armature must alternate.

Tissue paper in electrostatic fields

J. C. Siddons, Thornton Grammar School, Bradford

There are plenty of quantitative experiments concerned with electromagnetic fields, but there are so few concerned with electrostatic fields that a simple one, though not of much accuracy, might be of interest.

Place a small square sheet of tissue paper at the middle of the bottom plate of a parallel plate capacitor. An e.h.t. unit establishes a potential difference across the plates and a voltmeter measures this p.d. (Fig. 1). Increase the voltage across the plates: eventually the paper will rise. If the length of the side of paper square is greater than the separation of the plates, the square remains as shown at (*a*) in Fig. 2; if smaller, then the square may either simply stand up (*b*) or may commute between top and bottom plates (*c*). In the latter case the square remains hanging from the top plate for a few seconds before it falls off.

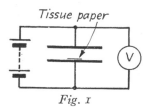

Fig. 1

Measure the lifting voltage (i.e. the voltage required for the paper to be lifted) for several successive lifts of the same piece of tissue paper. It will be found to be reasonably constant and reasonably predictable. Draughts or vibrations should be avoided: these will give lower than usual values for the lifting voltage. Sometimes on the other hand the voltage must be considerably above the usual voltage before lifting takes place, some kind of adhesion having taken place.

If thin metal foil is used in place of the tissue paper, lifting occurs, but because of the high conductivity, sparks begin to fly from the foil to the plates. This seems to produce adhesive effects and no consistency in lifting voltage can be obtained. The tissue paper works well simply because the paper is such a poor conductor.

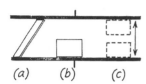

(a)   (b)   (c)

Fig. 2. (a) a piece of tissue paper stands up on the bottom plate and touches the top one. (b) a smaller paper square simply stands up. (c) another smaller square moves up and down between top and bottom plate

From the theory given below it would appear that the lifting voltage should not be affected by the size of the tissue paper, both electrostatic force upwards and gravitational force downwards being directly proportional to area and so cancelling out. However, testing shows that the size of the paper does have a definite though small effect. Rectangles of tissue paper, 2 cm × 1 cm, 1 cm × 1 cm, $\frac{1}{2}$ cm × $\frac{1}{2}$ cm, were found to require 1150, 1250, 1300 volts respectively. The explanation of this is the lack of flatness in the paper. However carefully the paper squares are handled they tend not to lie perfectly flat on the capacitor plate. The bigger the square the more the lack of flatness will show up: this means that part of the paper is nearer the top plate than it should be and so the paper is lifted up by a smaller voltage. Thus the smaller the tissue paper, the more will it behave in agreement with the theory.

A simple experiment is to find out how, for a particular square of tissue paper, the lifting voltage depends upon the separation of the plates. To separate the plates I used small perspex blocks of thickness 3 mm. These were stacked in piles. When there were 2, 3 and 4 blocks in each pile the lifting voltages were 1400, 2100 and 2900 respectively. These figures neatly confirm the important electrostatic fact that field strength depends upon potential difference/separation in the case of a parallel plate capacitor. The capacitor plates used had a diameter of 14 cm. With greater separations than those used it would be expected that the voltage would no longer be directly proportional to the separation.

The theory is as follows: the capacitance of the parallel plate capacitor is $\varepsilon_0 A/d$ where A is the area of either plate, d is the separation of the plates and ε_0 is the space constant. When the p.d. between the plates is V the quantity of electricity Q on either plate is $\varepsilon_0 AV/d$. Thus the quantity δQ of electricity on the tissue paper of area δA is

$$\frac{\varepsilon_0 V}{d}\delta A.$$

Let the field strength between the plates be E. One half of E is due to the top plate, the other to the bottom plate. Thus the force exerted by the field due to the top plate on an element of charge δQ on the bottom plate will be $\frac{1}{2}E\,\delta Q$. But $E = V/d$ and so the force δP on the tissue paper will be given by

$$\delta P = \tfrac{1}{2}E\,\delta Q = \tfrac{1}{2} \times \frac{V}{d} \times \frac{\varepsilon_0 V}{d} \times \delta A = \frac{\varepsilon_0}{2} \times \frac{V^2}{d^2} \times \delta A.$$

At the lifting point δP becomes equal to the force of gravity on the paper: if Δ is the mass per unit area of the paper, the pull of gravity is $\Delta g\,\delta A$.

Thus
$$\Delta g\,\delta A = \frac{\varepsilon_0 V^2 \delta A}{2d^2}$$

and so
$$V^2 = \frac{2\Delta g d^2}{\varepsilon_0}.$$

Knowing the mass per unit area of the tissue paper and the separation of the plates we can thus, purely electrostatically, calculate the voltage across the plates at the lifting point.

Six squares of tissue paper were cut with sides of 5 cm. They were found to have the same mass, 0.0465 g, giving for Δ a value of 1.86×10^{-2} kg m^{-2}. We cannot however assume that all squares of side 0.5 cm will have the same mass as each other and as this mass will be of the order of $\frac{1}{2}$ milligramme we cannot check by direct weighing. To test for uniformity of Δ I therefore measured the lifting voltage for several $\frac{1}{2}$-cm squares but found no significant variation in the voltage. Thus we can apply the value of Δ found for the bigger squares to the smaller ones.

Substituting 1.86×10^{-2} kg m^{-2} for Δ and 6.5×10^{-3} m for d, we find that the lifting voltage should be 1300. The voltmeter read 1400, showing fair agreement. Thus the experiment can be regarded as an electrostatic method involving the measurement of a force for checking the reading of a voltmeter. Alternatively we could make use of the experiment to measure the weight and so the mass of any small piece of tissue paper or similar material.

Electrostatic attractions and repulsions

R. W. Mortimer, King's College Junior School, Taunton

To demonstrate electrostatic attractions and repulsions, a conducting sphere was suspended on insulating thread near the charged sphere of a Van de Graaff generator. The expected attraction followed by repulsion is clearly shown; but the experiment became both spectacular and varied on placing an earthed sphere nearby. Points noted include:

1. Attraction of the charged suspended sphere to the earthed one and subsequent return for further charge.
2. If the uncharged suspended sphere is near enough to the earthed one on switching on, it moves the 'wrong' way—a useful source of discussion.
3. Occasional discharges through the suspended sphere as it passes between the other two.
4. A frequent movement round the earthed sphere with very close similarity to the orbiting of satellites. Discussion again here.
5. After switching off, movement back and forward between the main spheres carrying small proportions of the charge to earth.

One must be prepared, however, not to make any useful observations at all at the start of the lesson; the varied movements are far too spectacular!

Force-on-conductor balance

D. J. Wilkinson, Brudenell Mount, Leeds

The following is a variation of the 'force-on-conductor' balance used to investigate the relationship between the force on a current-carrying conductor, magnitude of current and the length of conductor in a uniform magnetic field.

The variable length conductor is firmly clamped, whilst the magnet providing the uniform magnetic field is placed on the pan of a sensitive compression type balance. The force on the conductor can then be measured as the reaction causes an equal measurable force on the balance. Magnadur magnets mounted on a soft steel yoke provide an adequate magnetic field.

If the yoke is rotated on the balance pan, the effect of angle between the magnetic field and the conductor upon the force can be investigated.

A flux linkage model

H. Greenstock, Haberdashers' Aske's School, Acton

A simple model showing the relationship between induced e.m.f.s and the number of turns linked by a flux can be shown as follows.

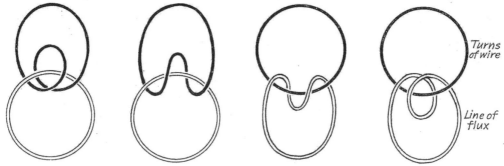

Turns of wire

Line of flux

Take two pieces of plastic coated wire of different colours and length approx 0.5 m. Join the ends of one in a circle, to represent a single line of flux. Make two turns of the second wire round the first and join it up to represent a couple of turns of a coil. The diagrams show how the wires can be rearranged to give a picture of two lines of flux linking a single turn of wire. As the actual linkage is unchanged, the same e.m.f. and current will therefore be produced and the theoretical significance of the number of turns is made clearer.

A little practice is advisable, to avoid loss of nerve in front of a class!

An application of the fickle ball bearing experiment

E. M. Royds-Jones, Wykeham House School, Fareham

At more than one lecture-demonstration I have seen Professor Laithwaite show that a ball bearing will leave a magnet to which it is attached and go in preference to a small piece of unmagnetized iron brought up to it. This phenomenon is useful if,

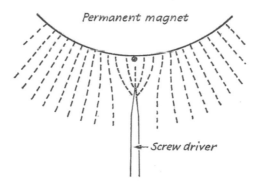

Permanent magnet

Screw driver

when dismantling or examining a meter, small pieces of steel wool or iron filings or other magnetic particles become inadvertently attached to the permanent magnet of the meter. It is difficult indeed to get rid of these by wiping the magnet, but they can be removed with ease by the tip of a small screw-driver. The diagram shows how the screw-driver causes a concentration of magnetic lines of force, indicating increased magnetic field to which the small magnetic particle migrates.

A direct measurement of charge

A. H. Stanway, Wolstanton Grammar School, Newcastle-under-Lyme

There is need in Section V of the Nuffield physical science course for methods of charge measurement which do not involve the use of a ballistic galvanometer. The method described below arose out of a discussion with pupils in which it was suggested that charge could be estimated from a current/time graph in the same way that distance is obtained from a velocity/time graph. This approach is obviously not original, and is not particularly accurate, but affords good practice in graphical methods and handling of electrical units.

Charge is stored by a capacitor and is then allowed to leak away through a resistor. The decay of voltage is monitored on an oscilloscope screen using the circuit shown in Figure 1.

A very slow sweep speed is required, a facility which is now available on educational instruments. The slowest speed on the Telequipment S51E is ideal for the purpose,

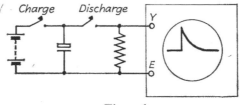

Figure 1

and the complete discharge can be accommodated if a 50 000 μF capacitor is charged to 12 V and then discharged through a 30 Ω resistor. If the capacitance is changed to 50 μF the resistor needs to be about 30 kΩ. Both voltage and time axes need of course to be calibrated. The sweep is slow enough to be timed by stopwatch and therefore presents no difficulty.

The best course now would be to photograph the trace, which I have not tried, but little practice is needed to transfer points from the oscilloscope screen to graph paper with fair accuracy. Ten discharges can be accomplished quite rapidly. Using the value of the discharge resistor, the voltage axis is transformed to a current axis (or another graph plotted), and the charge estimated by counting squares below the graph line.

I have found that the method is readily understood and works well as a class experiment using the Telequipment Junior oscilloscopes, and helps to clarify the concepts of charge and current. It also provides a very good basis for the subsequent discussion of the ballistic galvanometer.

Use of a toy motor to demonstrate rectifier action

W. H. Jarvis, Rannoch School, Perthshire

The 'toy' motors used in ripple tanks, the Malvern energy kit, etc., vividly illustrate the difference between a.c. and d.c., and the action of a rectifier, because they respond markedly to a.c.

A 2 V a.c. supply is sufficient. When the motor is connected directly to it, the spindle vibrates with considerable amplitude. A boss with a long fixing screw makes the 50 Hz vibration more obvious.

Almost any 1 A diode can then be put in series, first one way round and then reversed. The reversal of motor direction convinces pupils that the invisible 'current' has also been reversed.

Determination of the wavelength of u.h.f. TV transmissions

D. C. Gaskell, Matlock College of Education

The experiment can be carried out by any school where u.h.f. TV transmissions can be received, and it uses a straightforward interference method. It allows pupils to determine the wavelength of electromagnetic radiation without the need for specially designed equipment such as the 3 cm apparatus. Better results can be obtained in areas of weak signal strength.

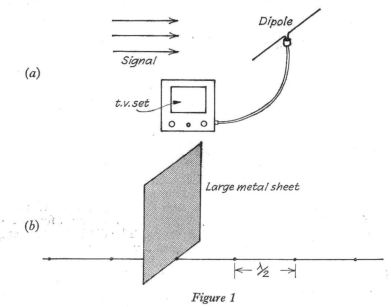

Figure 1

A small dipole, connected to a TV set by means of a coaxial cable, receives the signal from the transmitter, and the TV set is used to monitor the strength of the signal. If the signal is horizontally polarized, the dipole must be mounted horizontally with its length at right angles to the direction from the transmitter. If the signal is vertically polarized the dipole must be vertical. The polarity can usually be determined by trial and error or by observing the plane of aerials on houses. Adjust the position of the dipole in the laboratory to obtain a weak signal on BBC 1, BBC 2 or ITV. Place a large metal sheet (we used the tray from a laboratory trolley) vertically about 2 m behind the dipole away from the direction of the transmitter. Move it towards and away from the dipole and mark its position on the bench for each position of minimum signal, i.e. when the picture becomes unstable or disappears. It is not always possible to detect the minima if the direct signal is too strong. The distance between each minimum is half a wavelength. Move the metal sheet through a number of minima, measure the total distance moved and divide by the number of half wavelengths to obtain a mean value of $\lambda/2$ (Figure 1 (b)).

DETAILS OF DIPOLE CONSTRUCTION

The overall length of the dipole should be equal to half the wavelength of the signal. Frequencies of transmissions vary over the country but lie somewhere between 620

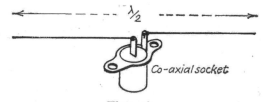

Figure 2

and 700 MHz (i.e. 48–43 cm wavelength). In strong signal areas the length of the dipole will not be critical but in weak signal areas it may be necessary to make the length as near to the half wavelength as possible.

Solder two lengths of stiff wire, each 12 cm long, to the tags on a coaxial socket. The length can be adjusted by cutting off small lengths from each side until a reasonable signal is achieved. Hold the socket with a retort stand and clamp.

The straw electroscope

Lalit Kishore, Mayo College, Ajmer, India

An electroscope can be made very simply as follows, and it works wonderfully. Take a cardboard box of the length greater than that of the drinking straw. Cut off the front and back face of the box except the two strips as shown below. Make a cut

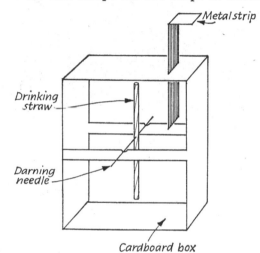

on each of the strips. Pass a darning needle through the straw about 0.2 cm away from its centre. Through the top face of the box pass a metal strip as shown. The electroscope is ready for use now. On bringing a charged body near the metal strip the straw shows a deflection. Furthermore, the electroscope can also be charged. Better results are obtained if the straw is painted with 'silver' paint.

Projection of magnetic field patterns

F. A. Randle, The High School for Girls, Northampton

This is a simple way of demonstrating to a whole class the field patterns around bar magnets and coils of wire and also the effects of the soft iron, etc. The source of the magnetic field is placed on top of an overhead projector and a sheet of acetate is placed over the top of it.

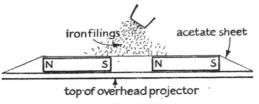

Figure 1

Iron filings sprinkled over the acetate sheet create an opaque field pattern which is then projected on to a screen. A plotting compass with glass windows on both sides placed on the acetate will give the field direction.

Light-emitting diodes

T. J. Ericson, City of Leeds and Carnegie College

These new devices which are used commercially for numerical display on calculating machines and digital meters, have uses in science teaching because of their low voltage and current requirements. For example the red light emitted by the diode Monsanto type MV 5025 (available from Integrex Ltd, PO Box 45, Derby DE1 1TW, at 35 p each) is visible with a forward current greater than 1 mA. The maximum forward current is 20 mA. The forward voltage at emission is 1.6 volts. The diode does not emit light at reverse bias and the reverse bias must not exceed 3 volts.

These properties of the diode have suggested the following experiments.

(*a*) Very simple and inexpensive cells can be constructed which will give sufficient current to light the diodes. An aluminium–copper cell can be constructed by wrapping one or two layers of newspaper around a piece of kitchen aluminium foil which has been folded a few times to form a firm rectangle of about 4 cm × 10 cm. The paper is held flat against the aluminium by winding about a metre of copper wire around the paper, and pressing the wire against the paper so that the distance between the wire and the aluminium is a minimum. Care must be taken that the wire does not touch the aluminium. The cell is completed by soaking the paper in dilute (0.5 M) sodium hydroxide solution, the copper being the positive pole and the aluminium the negative pole. Two of these cells in series will give sufficient voltage to light the diode. There is no need for cells to be immersed in the sodium hydroxide solution; the wet paper contains sufficient solution to light the diode for about 15 minutes.

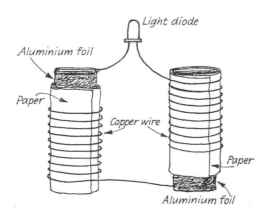

(*b*) An electrolytic capacitor of 10 000 μF charged to 25 volts will light the diode for about a minute, if the diode and capacitor are connected in series with a 4.7 kΩ resistor. In this experiment care must be taken not to reverse bias the diode.

Introduction of inductance by direct measurement

J. D. S. McMenemey, Rugby School

The most difficult concepts deserve the simplest introductions. Nothing could be more basic than connecting a pure inductor in series with an ammeter and battery (Figure 1), and showing the effect of V and t on the current in the defining equation $V = L\, dI/dt$.

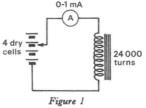

Figure 1

Unfortunately, two practical limitations of actual inductors complicate the issue, but by careful selection it is possible for them to nearly cancel out to give a most impressive demonstration of the phenomenon. Whereas resistance in the circuit progressively limits the growth of current, saturation of the core causes an increasing rate of growth.

The inductor used for these measurements is obtainable from Messrs Unilab Ltd. It consists of a large core of cross-section 1.3×10^{-3} m² (2 square inch) and 0.4 metre circumference fitted with two coils of $2 \times 6\,000$ turns each, and gives an inductance of about 50 kH. The graphs show the remarkable proportionality of the growth of current for 4.5 V.

The following demonstration procedure is very effective:

1. Connect to 3 V, and show how the current grows, and continues to grow, apparently uniformly (to at least full scale deflection).
2. Show the faster growth for 6 V; also the slower rise for 1.5 V, but avoid reaching resistance limitation.
3. For each of the four voltages, either time the growth to 0.5 mA or measure the current after, say, six seconds. It is neither easy nor fruitful to do this with great accuracy.
4. The defining equation can now be given and justified, and a rough measurement of L can be obtained.
5. By moving the top of the C-core, support is given to the belief that magnetic energy is stored when current flows. One of the coils can be reverse connected to show the effect of flux cancellation, which drastically reduces the time constant. Use of a higher range ammeter and the 6 V e.m.f. shows the core saturation occurring.

Two minor points should be mentioned by way of warning. Firstly, it is important that the air gap in the core is minimal; the two halves must mate as perfectly as possible. Secondly there is, even with a good core, some hysteresis effect. The graphs in

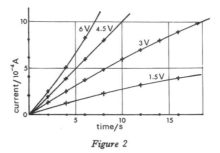

Figure 2

Figure 2 were obtained by breaking the magnetic circuit to demagnetize between successive runs. Even longer growth times are obtained if the current is reversed between runs.

A magnetic levitator

P. Violino, Istituto di Fisica, Università di Pisa, Italy

Kinsella [1] has recently suggested how to float a magnet, showing the well-known phenomenon of magnetic levitation. An easier and perhaps more impressive way to do it, is the following one. Obtain two ferrite magnets of the kind used to post notices on iron billboards. They are usually 3 mm ($\frac{1}{8}$ in) thick and are either square or round with a 2.5 to 5.0 cm (1 to 2 in) size. It is easy to obtain them with a hole in the middle, but if you can't,

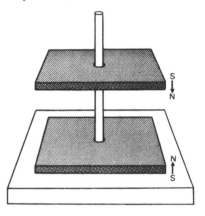

you can easily drill a hole in them. Instead of using the cage suggested by Kinsella, you can simply use any non-magnetic vertical stick (a toothpick will do) and place the two magnets around it with the like poles facing each other, as shown in the figure. The upper one will levitate and in addition will freely rotate around the central stick.

REFERENCE

1. Kinsella, D., 'The floating magnet', *S.S.R.*, 1980, 216, **61**, 546–7.

An electrostatic 'bell'

Alan Ward, Saint Mary's College, Cheltenham

Benjamin Franklin was so engrossed in his studies of electrostatics that he once gave an 'electrical picnic' on a bank of the river Schuylkill, which runs through Philadelphia, USA. According to Thelma Harrington Bell, in her book *Thunderstorms* (Dobson, 1963): 'Brandy was set alight, and a cooking fire was kindled, both by sparks from Leyden jars. Then a turkey for the dinner was dispatched on the spot by a severe electric shock.' Franklin even constructed an electrostatic 'bell' that told him when 'lightning clouds' passed over his house.

Such a bell is soon improvised. Stand each of two large bare-sided tins upon separate inverted polystyrene trays (to serve as insulators). Bridge the space between the tins, using a third tray. Then cover a ping-pong ball with aluminium foil, and dangle this object—by means of cotton and sticky tape—midway down between the tins. There should be a distance of two or three centimetres between the ball and the tins on either side. Electricity can be generated by rubbing a transparent plastic squash bottle with a woollen scarf.

Afterwards, the (negatively charged) bottle is brought up closely to an outer part of one tin. The metalled ball then starts to swing, and to strike the tins' walls audibly. How does it work? Electrons upon the bottle repel electrons from the atoms on the nearest part of the tin. These repelled electrons gather on the tin, next to the ball. At first the ball is attracted; but, on touching the tin, it collects extra electrons, and is consequently repelled towards the second tin. The ball gives up its surplus electrons to the other tin—and the cycle continues.

Figure 1

Franklin's device was (dangerously!) joined to a lightning conductor, down which electrons, repelled by the negatively charged underside of a thunder cloud, could travel to his bell. Simulate this by erecting a straightened wire coat-hanger on top of one of the tins, and by simply moving the charged bottle 'cloud' above its upper point. Juniors observing these experiments asked why the bell also rang just after the bottle was removed. This was because electrons which had gone across to the second tin began a reverse cycle of events, leading to electrical stability in the system.

Our most entertaining activity involved a child standing upon an upturned plastic (insulator) bowl. We charged the child by repeatedly rubbing the bottle, before discharging the accumulated electrons on to a hand, via outspread and upward-pointing fingertips. While this was happening, the child touched the nearest tin with the free hand—causing the bell to ring. An electroscope (made by dangling two pea-size cork balls, side by side, on the ends of a cotton loop) was held near the second tin. During this experiment, this simple instrument diverged, to demonstrate the transfer of electronic charge.

Like charges repel on an amazing electrified hat

Alan Ward, Saint Mary's College, Cheltenham

Fashion the traditional toy pointed hat from a full-size sheet of sugar paper (Figure 1). Cut off most of the point and turn the object upside down. Now use a stapling machine to clip the loose flaps at the front (Figure 2) to keep them secure. Then cut a series of 0.5 cm wide strips of coloured tissue paper, and fasten bunches of these bright 'tassels' to the outer projections of the modified headgear. If all is done quickly in

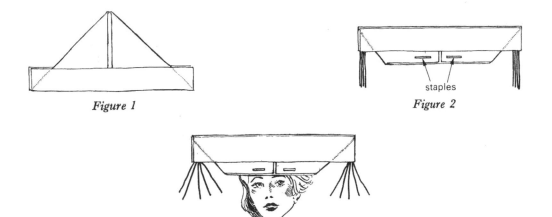

Figure 1

Figure 2

staples

Figure 3

front of the children, so much the better because the construction itself will stimulate their attention.

Invite a child to stand upon an inverted plastic tray which is reinforced beneath with books. The tray will insulate the child electrically from the ground. Next charge a clean dry plastic orange squash bottle with electrons, by vigorous rubbing with rough wool. Keep discharging the static electricity to the spread-apart fingertips of the volunteer. At the same time draw attention to the tassels on the hat. Everybody will be delighted to see how the 'like' charged paper strips repel one another and open out most strangely! (See Figure 3.)

Photoelectric effect

W. K. Mace, King Edward VII School, Sheffield

An essential element in presenting the case for the photon concept (Nuffield A, Units 5 and 10) is the following argument. On the basis of the assumption that light energy spreads out uniformly over an expanding spherical wavefront, it should take a calculable time for any single photocathode electron to collect enough energy to be emitted; in fact we find that even under feeble illumination photoemission starts without detectable delay.

This can in fact be demonstrated without trouble. Simple calculation shows that using a 2.5 V torch bulb at 1 m distance, and assuming a 1 to 2 per cent luminous efficiency, it should take any single surface atom several thousand seconds to absorb 1 eV of energy. However, the Unilab photocell shows an immediate response under these conditions.

The argument of course begs a lot of questions, but if one is to use it at all, seeing the thing happen is better than just talking about it.

Audible electric currents

W. K. Mace, King Edward VII School, Sheffield

The 'water circuit' was invented to try to fix in pupils' minds what we mean by current, yet, as we all know, some pupils still do not relate properly an ammeter reading to the rate of flow of charge. Even amongst students who can scrape an A-level pass there are some who suffer from this disability, and it causes particular trouble in the context of reactive circuits. However hard you try—using slow a.c., imaginary rubber-membrane capacitors, pulleys and springs, the double beam oscilloscope—the concept of *why* there is a phase difference between V and I continues to evade them.

It occurred to me that this mental blockage might be cleared if we could somehow find a means of '*hearing* the charges moving round the circuit'. A digital joulemeter might do it, but so far we have not bought one. Then a colleague suggested we might somehow use a motor, and we have made the thing work, with immediate success for one group of students. The arrangement described is specifically for the capacitor problem, but clearly it could be valuable at a wide variety of levels.

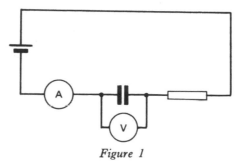

Figure 1

The circuit is shown in Figure 1, with the components laid out more or less as they would be on the bench. Because it looks rather frightening, it is essential that the students' attention is directed to a clear simplified circuit diagram (Figure 2), and that they should concentrate on the indicators on the bench and not on the wiring. The slow a.c. unit is hand operated, and ours is heavily geared down by means of a small Meccano pulley and string drive so that it can be worked very slowly and steadily. The motor is the 'mini motor' supplied with the Nuffield A semiconductor thermopile, and it is fitted with a small drinking straw propeller to show up its rotation. Since this motor is very quiet we put a 400 mW speaker-amplifier unit across it to amplify the brush noise. The ammeter is centre-zero, 2–0–2 mA, shunted as necessary with a rheostat; it stands behind the motor. For indicating the p.d. across the capacitor the small CRO is set to 'd.c.', and it is laid on its side so that the movement of the spot is like that of a meter pointer.

A suitable teaching sequence would be something on the following lines. First, short out the capacitor, and use the slow a.c. generator simply as a source of variable steady voltage applied to the resistor. When the voltage is changed or reversed, or the value of the resistor changed, we can watch the ammeter, and also watch and listen to the rotation of the motor; it is very easy to identify the noise as representing 'the sound of the coulombs going round the circuit'.

With the capacitor in circuit, the first thing to do is to play around with the applied voltage, 'hearing' current flowing and at the same time watching the p.d. as it changes. Charge can be put on, more added, some taken off, allowed to go on flowing so as to reverse the p.d., and so forth. The last thing one does here is to carefully move charge back and forth so as to get an exactly uncharged capacitor. This leads to the next step: exponential charging. A rapid 90° turn on the generator applies full voltage suddenly. Immediately we hear a large current as the capacitor begins to charge, and then it is heard to die away gradually as the capacitor gets nearer and nearer to its final p.d. We can discharge the same way—then we can increase the resistance and hear at once that the initial current is less, and the charging time longer.

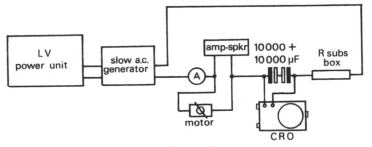

Figure 2

Finally comes the a.c., which should be at about 1/6 Hz or less so that there is plenty of time to see and hear what is happening. There is no doubt about this; as the p.d. begins to move away from maximum we hear the current rise, its greatest value is as the p.d. is passing through zero, and it dies away to nothing as the reverse charge accumulates. A glance at the propeller verifies that the next stage is current reversal. It is all very vivid, and the students referred to went off looking distinctly pleased and claiming that at last they could understand what it was all about. It might even be true.

An aid to teaching the d.c. electric motor

Randle Hurley, Avery Hill College of Education

Many students gain an understanding of the theory of the d.c. electric motor from the diagrams found in most textbooks but when the use of these fails to bring about understanding, it seems that the pupils are unable to grasp the interdependence of the several changes that are occurring at once. The changes in current direction, magnetic field and armature position cannot all be manipulated at once in order to appreciate the relationships. Small electric motors can help borderline pupils make the connections but fail to do the job for all pupils for two main reasons. First, the motors usually have a large, enveloping magnet which prevents the pupils seeing all the constructional detail and second, and more important, the motors cannot display changes in the magnetic field round the armature that occur as the motor rotates. I designed the model motor described below to overcome the shortcomings of conventional models. The new model does the job of the usual board diagrams but, because it is infinitely variable, it allows the diagrams to be merged into one another. The state of the electromagnetic field surrounding the armature is displayed at all times and the pupils find it an easy task to decide how the motor will behave as the permanent and electromagnetic fields interact. Handling the model and watching the changes as they occur helps in the internalizing of the concepts by giving scope for enactive and iconic representation. This benefit is felt by pupils of all abilities, those who 'understand' first time as well as those who have difficulty with the subject.

However, some pupils, those who may go on to O- and A-level exams in physics,

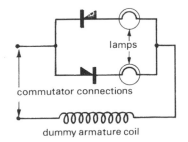

Figure 1

may have to 'unlearn' the concept gained at this stage in order to appreciate that motors without easily recognizable poles work because of the force acting on any conductor, not necessarily a coil, in a magnetic field. In the light of this it may be felt that a different style of presentation is necessary for some ability groups.

The extra information about the magnetic state of the armature is displayed by means of a pair of lamps and a pair of diodes arranged as in Figure 1. As the current flows in the first direction, determined by the position of the commutator relative to the brushes, one lamp illuminates the appropriate symbols on the faces of the armature. Turning the crank through 180° causes the current to flow the other way, energizing the second lamp and changing the display. The armature coil need only consist of a couple of turns. The spacing should be sufficient to allow the pupils to decide easily the direction of current flow. At some point on the armature the wire has to enter the hollow body to energize the circuitry. It would be better if the break in the coil were invisible and this can be achieved by the use of a small cable clip.

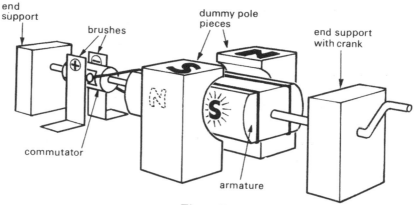

Figure 2

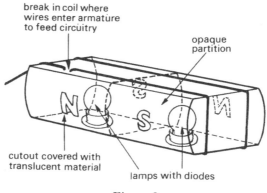

Figure 3

In the prototype the permanent magnet was represented by two suitably labelled blocks. There is a danger here that the existence of monopoles may be implied. I have tried to avoid this problem by introducing a normal, working model at first and, subsequently, introducing the new model with the explanation that the bulk of the magnet has been cut away for clarity. It may be felt that a better, if bulky, solution would be to construct a dummy, removable magnet which leaves the labelled poles as shown in Figure 2 on removal.

My prototype (Figure 3) was made from scrap materials. The diodes and lamps are available in most school laboratories. Any diode which will take the current drawn by the lamps will do. Power can be supplied by battery or the normal lab pack. The operating voltage is determined by the impedance of the diodes and can be found by gradually increasing the voltage until the lamps are reasonably bright. The brushes were made of card faced with cooking foil and power was fed to them with crocodile clips. Tension was applied to the brushes by means of a light elastic band. The commutator was made from a piece of broom handle with foil contacts stuck on. The coil leads were attached to the commutator with drawing pins. The armature was made of fairly thick balsa wood and card. A coat of matt black paint helped with light proofing and after painting greaseproof paper was stuck behind the cutout symbols.

Five minutes' fun with a silicon-controlled rectifier (thyristor)

John Marshall, Christ College, Brecon

The following experiment is so easy to set up and demonstrate, and such good entertainment that it is a pity to miss it out when a sixth form is recapping the fifth form work on half-wave rectification of a.c.

The silicon-controlled rectifier (scr) is put in series with a suitable light such as a 12 V 24 W bulb across a 12 V 50 Hz supply. An audio oscillator is now used to impose an a.f. signal between the cathode and gate of the scr, and its frequency is adjusted to 49 Hz. The light now goes on and off once a second, and the cro trace of the voltage across the lamp shows the gating action move across the sinusoidal pulse once a second, as in the diagram. A double beam oscilloscope is a help here; the Y_2 beam can be made to show the 49 Hz a.c. signal drifting across the screen at the same rate as the vertical part of the Y_1 trace where the scr suddenly starts to conduct.

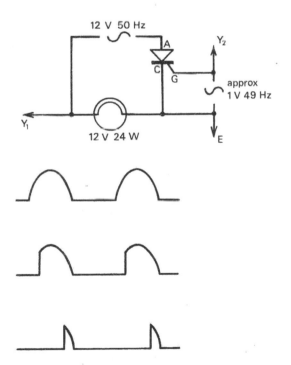

Maybe it isn't in the syllabus, and maybe they don't understand it, but they come back in crowds to try it for themselves, and they never did that with a straight half-wave rectifier.

The scr used was a surplus unit available from The Surplus Buying Agency, Richmond College Annexe, Station Road, Woodhouse, Sheffield S13 7RD.

Fascinating electrostatics

Alan Ward, Saint Mary's College, Cheltenham

Staple some strips of coloured tissue paper around the edge of an inverted metal foil tart case, to make a model flower having dangling petals. The paper strips can be less than a centimetre wide and about 12 cm long. Fix the model, using Blu-Tack, on top of a wooden stalk made with a dowel rod which is approximately 40 cm tall. The stalk should be inserted in a wood block base, and be standing upon an upturned plastic ice cream container, serving as an electrical insulator.

Charge the surface of a clean dry orange-squash bottle with static electricity, by vigorous friction, using a rough woollen scarf. Electrons rubbed off the scarf give the bottle a negative charge. When the charged bottle is supported high above the flower, and gradually brought down closer to it, beautiful effects are produced when the petals are attracted upward, and can be made to sway in an invisible electric 'breeze'. The so-called breeze is the result of moving the bottle's electric force field to and fro.

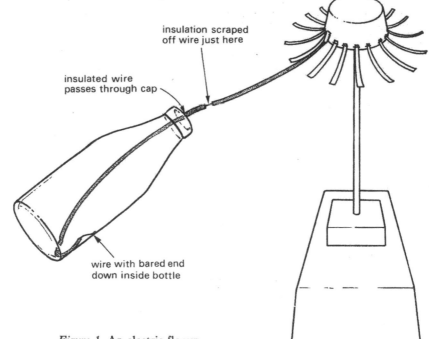

insulation scraped
off wire just here

insulated wire
passes through cap

wire with bared end
down inside bottle

Figure 1. An electric flower

Attach 2 or 3 m of insulated bell-wire to the flower, using a crocodile clip. Lead the bared opposite end of the wire *tightly* through a tiny hole bored in the cap of the bottle. The wire must go well down inside when the cap is screwed on. It should now be possible to demonstrate the connection between static and current electricity. Holding while rubbing the bottle, with the wire kept off the floor, produces a dramatic 'blooming' of the flower (Figure 1). Electrons on the bottle repel electrons along the wire.

The current 'flow of electrons' along the wire causes electrons to concentrate upon the flower, resulting in repulsion between its now negatively-charged petals. If, 20 cm in front of the cap, insulation has been scraped off the wire, electrons on the flower may be discharged by touching there, so collapsing the petals. Of course, for the most spectacular effects, these experiments must be done in a dry still atmosphere, such as indoors on a cold and frosty day. Beware of humidity. (Hint: After rubbing the bottle, hold the scarf away from it.)

While the bottle is resting alone upon the ice cream box insulator, the charged container can be attracted toward the palm of the hand. By moving the hand at critical moments as the bottle rocks back and forth, resonance can be shown. The bottle is made to rock and roll more and more, until it falls off the improvised platform. The charged bottle may also be used to attract tissue paper butterflies (Figure 2), or a floating toy duck, and—most amusingly—'electric caterpillars' (S-shaped polystyrene packaging modules).

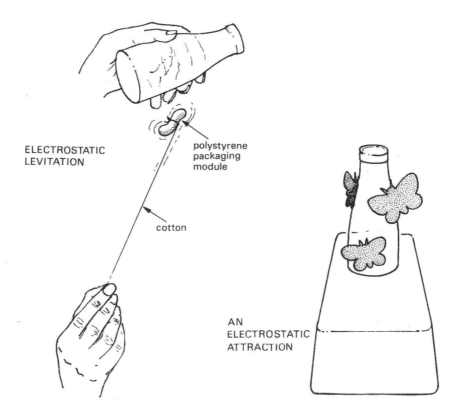

ELECTROSTATIC
LEVITATION

polystyrene
packaging
module

cotton

AN
ELECTROSTATIC
ATTRACTION

Figure 2. Pupils activities in electrostatics.

During my investigations, the charged bottle was at one time put down next to the 'stalk', just below the flower head, upon the ice cream box. At once the petals began to open out. Evidently the negative charge on the bottle repelled electrons in the paper, inducing positive charges on the ends of the petals. These positive charges repelled one another and opened the flower, like the diverging leaves on an electroscope. Holding the point of a pin just above the metal centre caused the petals to droop again.

Electrons must have escaped from the flower across to the pin and through my body. This would no doubt make the induced positive charge upon the petals much stronger, causing an effective field of attraction between the bottle and petals. The explanation seemed likely when, on my removing the bottle, the (positively charged) petals opened again. The flower could be opened and closed by putting the bottle back and removing it. While, in this condition, the petals were open, I neutralized the charge on the flower by touching its metal centre. The petals collapsed and hung still.

Firing 'Volta's pistol' with muscle-generated electricity

Alan Ward, College of Saint Paul and Saint Mary, Cheltenham

An effective Leyden jar, for storing electric charge, can be made by lining the lower half—inside and out—of a topless plastic bottle, using kitchen aluminium foil. The top of the bottle is replaced, and from it dangled a long loose chain of metal paperclips, to fall inside the bottle, and to rest upon the inner foil layer. Where the top link of the chain is attached to the bottle neck, an end of the paperclip is bent upward, forming a spike.

To charge the 'jar', vigorously rub a piece of plastic pipe, or a plastic orange squash bottle, with an old woollen scarf. Electrons collect upon the plastic material. Then move the plastic past the spike, allowing electrons to crackle down on to the chain, to be

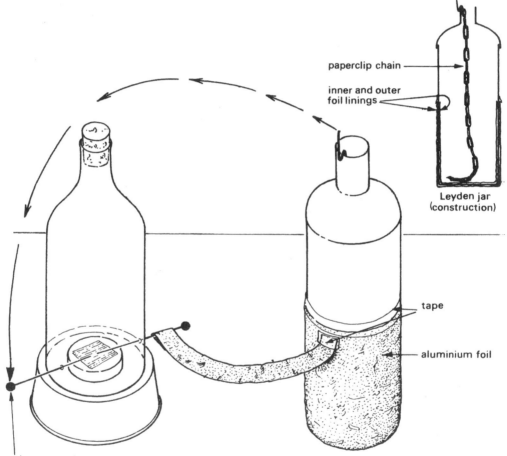

paperclip chain

inner and outer foil linings

Leyden jar (construction)

tape

aluminium foil

make contact here

concentrated on the inner foil lining. (If we imagine the charged bottle to be a cloud, the jar is very like a church tower with an earthed lightning conductor.)

Force exerted by the electrons inside the jar repels electrons from the outer foil lining 'to earth', which can simply be the table upon which the jar is resting. Thus the outer foil acquires a positive charge (a shortage of electrons and, therefore, a surplus of protons) by induction. In this way, the layers of foil become electrically polarized. Repeated discharges from the plastic pipe concentrate energy inside the 'storage' jar.

Opposite charges on the bottle can be neutralized by bridging the gap between the outer lining and the top clip point, using a finger and thumb. (This can be painful—so don't make the test with a heavily charged jar!) Now the jar can be used to demonstrate a device invented by Volta in the eighteenth century: 'Volta's pistol'. At the time it was only a sophisticated toy. As we shall see, the modern version of it has transformed our lives.

Construct the 'pistol' by cutting the bottom from a transparent plastic orange squash bottle. Two centimetres above the open bottom, insert the points of a pair of hat pins, to *almost* meet in the middle of the bottle. Stand the bottle upon an inverted plastic margarine tub. Then connect one of the pins to the outside of the Leyden jar, using a 2 cm wide strip of aluminium foil. Tape the foil strip to the jar's outer foil lining.

The well-charged Leyden jar can now be discharged by touching its point against the second hat pin. (Don't touch any foil while doing this.) A blue spark should flash inside the bottle. Make sure that all is working well, before putting a small jar-lid under the spark-gap inside the bottle. The lid should contain a fragment of paper towelling, soaked with several drops of petrol from a handy can of lighter-fuel. Drop a cork to fit loosely into the bottle neck.

Charge the jar well. Put down the plastic pipe and scarf. Hold the bottle against its pedestal with one hand, while using the free hand to discharge the jar against the second hat pin. If all goes well, heat from the spark ignites the petrol vapour and air mixture—causing an explosion of flaming yellow gas, which expels the cork with a 'pop'. The flame suffocates itself instantly, and safely. Yes, we have a model of the internal combustion engine.

A simple battery holder

D. J. Daniels and J. Snaddon, formerly University of the Witwatersrand, South Africa

The holder as illustrated can be easily constructed from oddments and has been found to give reliable service.

Take a 20 cm length of PVC water pipe of 35 mm diameter and heat one end in boiling water. It is then possible to screw a lid from a 62.5 g Marmite jar or one of similar size on to the softened end. If the end is now plunged into cold water a suitable thread will be formed. Drill a hole through a metal disc (or coin) of about 15 mm diameter, pass the end of a 30 cm length of electrical wire through it and solder it into place and then thread the wire through three 1 cm thick polyurethane foam discs cut to fit into the lid, which can now be screwed on to the end of the pipe.

Three 1.5 V batteries are now placed in the pipe with their negative end directed towards the lid; press them in firmly and then drill two holes through the piping and insert a nail to hold the batteries in place. A piece of wire can then be soldered to the nail and construction completed by attaching crocodile clips to the free ends of the two wires.

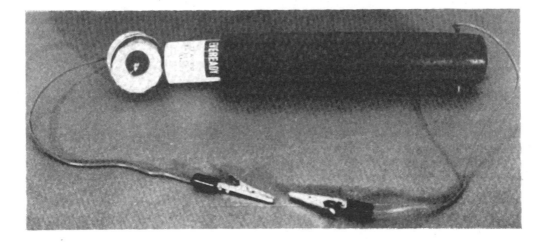

The use of the top pan balance in physics

D. K. Lomas, Bradfield College

Measurement of small forces reliably in schools has never been easy. The advent of the electronic top pan balance has changed this. Simple experiments in both electricity and magnetism have become reproducible with Nuffield A- and O-level apparatus.

Here are a few of the many experiments that can be attempted:

polythene rod

Figure 1. Electric lines of force

(a) Capacity of a parallel plate capacitor.
(b) The force of attraction of the plates of a parallel plate capacitor.
(c) The electric field between the plates of a capacitor $E = F/q$.
(d) The force between charges. Inverse square lens.
(e) The flux density of bar magnets, coils, wires, etc.
(f) Verification of $F = BIL$.
(g) Measurement of ϵ_0.

Typical reading at 1 cm from balance

$$\sim 4\,\text{g} \sim 4 \times 10^{-2}\ \text{N}$$

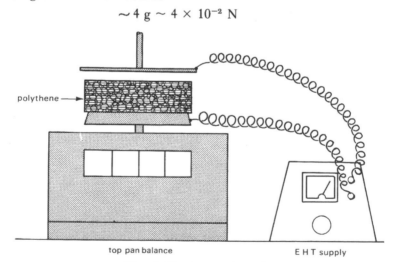

polythene ⟶

top pan balance   E H T supply

Figure 2. Measurement of ϵ_0 and ϵ_r

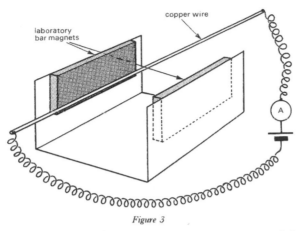

Figure 3

A parallel plate capacitor is formed between the top pan (metal) and the top plate (Nuffield Item 65) see Figure 2. If the battery is kept connected then theory gives:

$$\epsilon_r = 2Fd^2/V^2\epsilon_0 A$$

Typical results:

V = 3000 V
F = force of attraction measured = 6.50×10^{-2}N
d = 3.00 mm
A = 128.6 cm²
Hence $\epsilon_r = 1.14$

The balance is equally as versatile in magnetism. The measurement of the B field of a bar magnet is shown and this is given directly from F = BIL (see Figure 3).

A demonstration in electrostatics

C. G. Hanson, Mid-Herts College of Further Education, Welwyn Garden City

While recently discussing the transfer of electrons during a lesson on static electricity and the charging of insulators by friction, the following demonstration emerged.

When an alkathene rod is rubbed on the teacher's head (more interesting to the class than using the traditional cat-skin generally complete with tail), although it is perfectly easy to show that the alkathene becomes charged negatively, any attempts to show that the hair/fur has been charged positively fail because of the discharge-at-points phenomenon.

When, however, the alkathene rod is rubbed with an inflated balloon, then both rod and balloon have the desired effect when brought up in turn to the gold leaf electroscope.

The beauty of the demonstration now follows: when both rod and balloon are brought down towards the uncharged gold leaf electroscope together, the gold leaf remains in its inert state, while, by raising or lowering either rod or balloon, the leaf rises and falls at will. Raising or lowering both rod and balloon together causes no change in the state of the gold leaf.

The fact that positive and negative charges can cancel each other out now becomes clear; also that no matter how much negative charge is transferred to the alkathene rod, an identical positive charge remains on the balloon.

Electric writing

John Mattocks, Bristol Polytechnic

The following experiment is fun, easy to do and instructive.

Soak a piece of filter paper, or white blotting paper, in potassium iodide solution. Remove the paper and shake off the excess solution leaving the paper damp, but not soggy. Spread it out on a sheet of tin plate such as a large tin lid. Connect the lid to one terminal of the *low tension* supply of a lab pack and a *defunct* ballpoint pen to the other. If the ink tube is removed from the pen, a thin copper wire can be threaded down the barrel and can be nipped between the case and the brass end. Now write on the paper with the lab pack switched on to 6 V d.c. Try again with the polarity reversed and again on 6 V a.c. (avoid touching the tin with the pen point to prevent blowing a fuse).

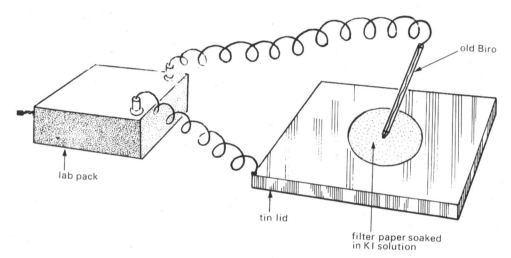

old Biro

lab pack

tin lid

filter paper soaked in K I solution

Figure 1

There is a number of questions that can be asked such as:

What makes the brown line appear?
Why does a mark appear with one polarity and not the other?
 (The answer to this is not as simple as it at first appears to be!)
Why is the a.c. line dotted?
What does the interval between the dots represent in time?
Are the dots and intervals of equal length? Why?
How could you use the principle to measure the mains frequency and, given the frequency,
 to measure the revolution speed of a record player turntable?
Why do the brown marks fade after a while?

Keepers on the eclipse major

M. D. Ellse, St Paul's Girls' School, London

I have suffered various minor injuries and inconvenience trying to remove the keepers from our Eclipse Major magnet. Pliers are not always to hand and even sliding the keepers off is impossible for many pupils. Recently I drilled a hole through each keeper and screwed them to a piece of wood. That seems to have solved the problem.

The force on a current carrying conductor: a demonstration

D. L. Thompson, De La Salle College, Manchester

An interesting demonstration of the force on a current carrying conductor in a magnetic field can be presented with the apparatus shown in the Figure.

The lamp used was a Mazda 240 V 100 W clear 'coiled coil' type. The lens is converging with a focal length of approximately 30 cm (not critical) and any large, white surface will serve as the screen.

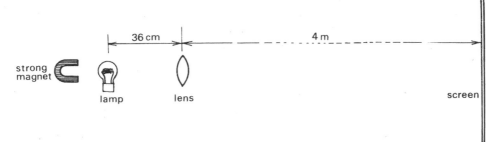

The lamp is switched on and a magnified image of the filament is focussed on the screen. The focus is sharp and some experiment may be needed with the distances indicated in the Figure if other optical arrangements are used. As the magnet is placed alongside the lamp, the filament can be seen performing quite large oscillations.

With the lamp suggested, the demonstration can be presented without a blackout.

The versorium: a resurrected static electricity detector

D. P. Newton, Spennymoor Comprehensive School, Durham

The *versorium*, shown in Figure 1, was invented at the end of the sixteenth century by William Gilbert, long before the better known leaf electroscope. As a teaching aid in elementary science, its potential does not appear to have been realized. It is inexpensive, effective and easy to make. In its simplest form, an inverted V of aluminium foil is

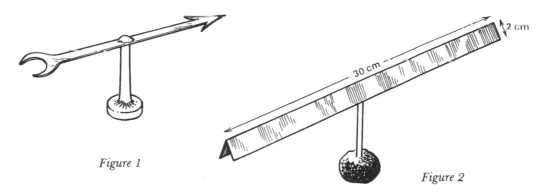

Figure 1

Figure 2

balanced on a long pin, as in Figure 2. The foil turns readily towards a charged body. To charge the foil itself, and to permit sign-of-charge testing, the device is stood on a square of plastic.

Children are impressed by the size of the detector and the alacrity of the movement. They can easily make one for their own practical work.

Control and measure of battery operated devices

Michael J. Williams, Barstable School, Basildon, Essex

It is convenient to be able to turn a tape-recorder or radio or toy vehicle on or off without taking it apart and soldering. It is useful to be able to measure current flow without cutting leads. This simple trick allows both operations without any alterations.

Figure 1 shows the idea. Two discs separated by an insulator are placed between any adjacent cells. Wires from these discs lead to an ammeter or to a switch. The latter could be relay operated or linked to a micro.

In practice I use a much simpler technique Figure 2. A length of plastic sticky tape holds two foil strips with wires twisted onto the foil. The foil is folded back on itself to act as the insulator. The assembly is minute and will fit into the smallest personal recorder, the thickness is negligible.

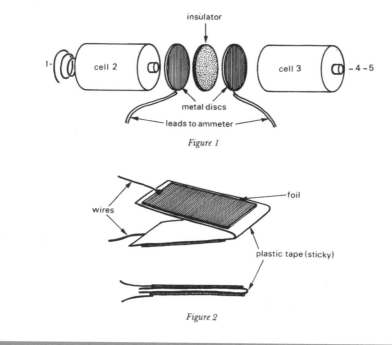

Figure 1

Figure 2

Coulomb's law using a top pan balance

Jonathan Osborne, North London Science Centre

One of the major problems of teaching physics is the problem of demonstrating sophisticated ideas convincingly. It is this that has lead to that well known aphorism that if 'Its physics, it doesn't work'. Such an idea is Coulomb's law which is notoriously difficult to demonstrate due to the usual problems with electrostatics experiments. The arrival of sophisticated and accurate digital balances has enabled a more accurate and careful demonstration of some of these principles. This experiment is an alternative to the traditional experiment outlined in [1].

The apparatus is assembled as shown. The polystyrene balls are coated in a conductor (either silver paint or aquadag) and stuck to the glass rods. One of the glass rods is attached to a polythene base, readily available from an electrostatics kit, either with glue

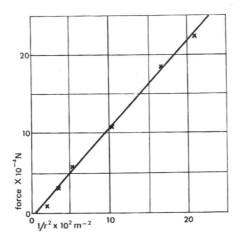

or Plasticine and placed on the balance. The balance must be capable of reading to 0.01 g. The other rod is supported vertically from a retort stand. The lamp is used to back project the positions of the balls onto a screen which has a sheet of graph paper on it.

The balance is zeroed with the tare and set to its most sensitive range. The balls are then charged with a proof plane which has been charged by induction. It is essential that this method is used as an EHT supply does not charge the balls sufficiently. The reading on the balance is then taken and the positions of the balls marked on the graph paper. The top rod is lowered and a new reading taken. This is repeated until sufficient readings for a graph have been taken.

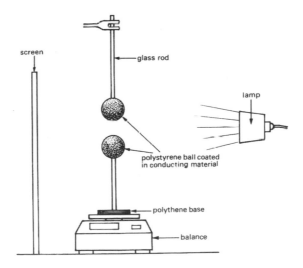

As can be seen, quite reasonable results are obtained and the method is very simple in essence. It does not require any complicated resolution of force in order to obtain a value for the force as its value is read directly. A possible reason for the values which have a low value for r not lying on an exact straight line, is that the approximation that the charge behaves as if it is at the centre of the ball is no longer true, as the charges are being pushed to the extreme ends of the ball. The readings that have high values of r, and hence low values of $1/r^2$, suffer from the fact that a reading can only just be taken on the balance with a consequent high percentage error in the reading. There is no obvious reason why the graph does not go through the origin.

REFERENCE

1. Nuffield, 'A' Physics Teacher's Guide, Unit 3, Expt 3.13.

Using LEDs to demonstrate induced current

C Lopez and P Gonzalo

There are many experiments that can be used to show plainly the existence of induced current in the study of Faraday's law.

An ammeter with a moving needle attached to a coil is the instrument usually used in such experiments, but it is not very practical in a classroom situation as it cannot be seen by many students at a time.

Electroluminescence solves the problem of demonstrating the existence of induced currents, due to the direct conversion of electric energy to light energy.

One type of electroluminescence device is the light emitting diode (LED), which is a semiconductor that radiates light from its PN junction. Depending upon the type of diode used, the supply voltage needed varies between a minimum of 1.5 V and a maximum of 5V.

Two LEDs, one red and other green, a coil and a straight magnet are the basic elements used for the demonstration. First of all, the LEDs are connected parallel to one another as shown in the figure. After testing the experiment with a coils set, check to see if a coil of 2000 turns is enough to produce the desired effect. The least number of turns needed to detect the induced currents is 500, but depends on the speed with which the magnet moves with respect to the coil.

The room should be darkened before beginning the experiment. The magnet is then placed in the hole of the coil and is then pulled out rapidly; a flash of red or green light is observed, depending on the magnet pole introduced. This shows that opposite currents are produced. The result obtained with LEDs of high brightness is surprising.

Thus the direction of induced emfs are specified and Lenz's law is verified, since the induced current appears in such way as to oppose the cause that produces it.

By using a coil of 2000 turns and varying the type of diode (depending on power supply voltage), the induced emf produced in each experiment can be approximately estimated.

C Lopez and P Gonzalo, IB Lope de Vega, Madrid

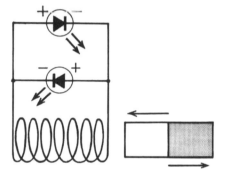

Figure   Basic circuit for induced current

The use of light emitting diodes to illustrate the nature of a.c. in comparison with d.c.

C. A. Bunker, University of Papua New Guinea

Modern cheap light emitting diodes may be used in junior work on electricity, where they can be used as 'bulbs' which are only able to work when the 'electricity' flows in one particular direction through them. By connecting diodes in series, it is easy to show that when the diode is off, it is in fact a non-conductor. Modern diodes do seem to be very rugged, and can be used with 2 V supplies without having to be protected with series resistors.

If a pair of diodes (preferably of different colours) are connected in parallel, but opposite ways about, it is easy to show that when connected to a dc supply, of two volts, from for example a laboratory power pack, one diode is on while the other is off Interchanging the leads to the diodes will illuminate the diode which was formerly off and turn off the one that was formerly on.

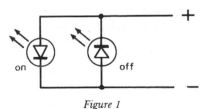

Figure 1

If the two diodes are connected instead to the ac terminals, *both* diodes illuminate. It looks as if ac goes both ways at once!

The situation becomes clear at once, however, if the two diodes are 'waved about' briskly with an amplitude of 30 cm or so. Due to persistance of vision, the 'on' diode will seem to leave a coloured 'streak' behind it, while the 'off' one will leave no such streak.

Figure 2

Subdued lighting is required, but the room does not need to be 'blacked out', and as long as direct sunlight is avoided, the experiment will work well. The diodes leave streaks which make it plain that one diode is 'on' while the other is 'off', and that they rapidly interchange.

The experiment may be considered a simple replacement for the old one where two probes were used to electrolyse potassium iodide/starch solution contained in a piece of blotting paper, leaving a similar trace in blue on the paper. The old experiment leaves a semi-permanent record, but is tricky to set up, and depends upon an understanding of electrolysis, whereas the new demonstration only requires that the pupils be aware that light emitting diodes are only 'on' for one direction of voltage across them.

Introducing electricity—ten projects for 11–12 year olds

Alan Ward, College of Saint Paul and Saint Mary, Cheltenham

THESE PROJECTS ARE SAFE

No illustrations are given because the aim of the exercises is to get children to approach the projects as problems, to be solved—as much as possible—in original ways. Your LEA Science Advisor will be able to suggest appropriate sources of bulbs, batteries, bulb-holders, various types of conductor wires—and some books, for reference.

1. Find as many ways as you can to light a 1.5 V (or a 1.25 V) torch bulb with a 1.5 V battery cell. But, before doing this, divide a paper into two parts. Call the parts 'Will light' and 'Will not light'. Then—as you do your tests—draw pictures of your ideas that work in the part called 'Will light' and pictures of ideas that do not work in the part called 'Will not light'. *In addition to the bulb and battery cell, you are allowed to use a narrow strip of aluminium kitchen foil.*

2. Light a 1.5 V (or a 1.25 V) torch bulb with a 1.5 V battery cell, by not using ordinary electrical wires, but by inventing circuits made from any materials or objects you wish. You could use a coat-hanger, scissors, paper clips, coins, keys, metal knives and forks, or a 'tin' can. Also try incorporating a little piece of coke, or a pencil lead. Make labelled drawings of all your electrical circuits that work.

3. Using a 4.5 V battery and a 3.5 V torch bulb and some single-strand electrical wire (with its insulation scraped off at both ends), invent some apparatus with which to test whether materials are conductors (that easily 'carry' an electric current) or insulators (that effectively prevent electrical conduction). Draw a labelled picture of your idea and make a chart of your tests on conduction and insulation.

4. Build a model bulb, using a screwtop jar and some Plasticine (to hold the filament support wires). The filament (a tiny spiral that actually lights up) can be made by winding very thin nichrome, iron or constantan wire around a pin. Test the model in a circuit with a 4.5 V battery and a homemade switch. Make a report on your project. Comment on your technical difficulties—and remark on how much light and heat your 'bulb' produces.

5. Inside a shoe box, join wires (bared at both ends) to paper-fasteners fixed to opposite sides. Number all connections on the outside of the box—and be sure to record correct 'combinations' before Sellotaping the lid on the box. (In some cases more than two paper-fasteners can be connected together.) Challenge other people to discover the combinations, without opening the box. They will use a 'tester' made from a bulb and battery and wires, from which go two bare-ended 'tester wires'.

6. What is a 4.5 V battery like inside? Carefully remove the cardboard casing from a *used* battery. Notice how the cells are connected inside. Draw what you see. Use a hacksaw and pliers to cut and tear open one of the cells. *Do this messy job over newspaper*. Again, draw what you see. Find your own words to label the parts. Look in a textbook to find out how the battery worked. Then write your own description of this.

7. Put long or short, thick or thin pieces of copper, iron, constantan or nichrome wires into a circuit planned to light a 3.5 V torch bulb (in a holder) with a 4.5 V battery. Test each wire separately. Then apply your experience to invent a way to make a light get brighter or dimmer, as you wish. Make an illustrated report. Include an explanation of why the bulb does not always glow so brightly (write your guess first—then look up the explanation in a textbook and write your version of what it says).

8. Invent your own bulb-holders and switch-operated lights. Use everyday materials like clothes-pegs, Sellotape, rubber bands, paper clips, aluminium foil, Plasticine and drawing pins. *It is important to make sure that all electrical connections between bared metal parts are firm and tight*. Your inventions are not much use if they fail to work after a good shaking . . . Make labelled drawings of your successes.

9. Find different ways to make one 4.5 V battery light *two* 3.5 V bulbs. You will, of course, need bulb-holders and wires. The bulbs will not always glow with the same brightness. Try to explain this in your own words—before looking for the answer in a textbook and rewriting your explanation. Make a report on your successes—and do not forget to mention how brightly the bulbs glow.

10. Against a white background, look at torch bulbs through a magnifying glass. Find out the meanings of numbers and letters stamped upon them. Draw a big cut-down-through diagram of a bulb. (This will mean taking a bulb apart *carefully*, by squeezing the casing with pliers.) Label the drawing in your own words. Show the path taken by electricity by marking your diagram with red crayon.

Electric penguins, conjuring tricks — and the paranormal

Alan Ward, Formerly of the College of Saint Paul and Saint Mary, Cheltenham

Plastic penguins, cut out from the wrappers of Penguin chocolate biscuits, can be electrified by stroking them between dry fingers. Presumably they collect free electrons from the body, which is an electrical conductor. If an electrified penguin is held up by one hand, it will be attracted by a finger of the opposite hand. Its negative charge repels electrons from the fingertip, to induce an opposite (positive) charge there. Then attraction takes place between opposite charges.

An electrified penguin will repel another one that is also charged in the same way (negatively). Or it will be attracted by a chrysanthemum flower—by induction; or it will 'stick' inductively to a window pane, and remain there for a long time. An electrified penguin is also repelled by a plastic drinking straw that is also electrified—by stroking along its length with a dry thumb and fingers.

Magical effects are possible with electrified plastic straws. For example, one will cling to the bottom side of outspread fingers on a down-turned palm of a hand. Coniurors use threads, adhesives, or the concealed finger from a hand gripping the wrist, to achieve similar 'magic'.

Rest such a straw across between the raised sides of a polystyrene food tray (from a supermarket). Hold an electrified straw in each hand. Then use the powerful force fields to repel the other charged straw to and fro along the polystyrene 'track'. It's unbelievable— and looks like certain effects said to be evidence for human psychokinesis (mind over matter), although the similarity with alleged paranormal phenomena should only put us on our guard when we are prepared to think about the actual occurrence of 'psychic' happenings.

An electric motor model that really works!

G. D. Giffould, College of External Studies, Konedobu, Papua New Guinea

I read about this motor [1] and have found it marvellous! Students make it easily, without all the fiddle of commutator, armature and axle, and bits of a Westminster Electricity Kit that get lost and have to be carefully assembled.

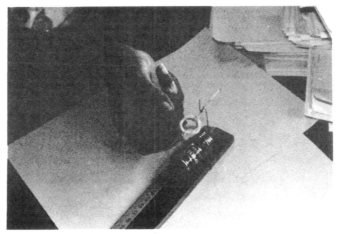

Figure 1

The motor does not need much to make:

1 m insulated copper wire, single strand
2 paper clips
rubber band or adhesive tape
1.5 V cell
1 or 2 magnets
cyclindrical former, Ø 10–15 mm, round which
to wind the wire e.g. test tube, finger, or AA size cell

The motor is easy to build:

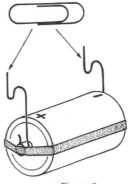

Figure 2

1. Wind the wire round the test tube or other former, leaving about 50 mm at either end.
2. Remove the former.
3. Loop each free end of wire once through the coil to stop it unwinding—see Figure 1.
4. Remove the insulation from each end of the wire.
5. Bend the paper clips as shown in Figure 2.
6. Hold the paper clips on the ends of the cell as shown, with the rubber band or tape.
7. Wedge the cell with objects at each side, to stop it rolling, Figure 3.
8. Rest the coil over the paper clips as shown. The coil must be level.
9. Hold a magnetic pole near the coil and spin the coil. It should continue to spin as an electric motor.

NOTES

The wire coil/armature must be light, but not so thin that the coil sags between the two paper clips.

Do not leave the coil on the clips without operating the motor. Students can try fixing a paper wheel or fan to the armature.

This simple motor project can be used to introduce the motor. The need for brushes, commutator, robust armature, etc. can then be developed for practical motors.

REFERENCE

1. *Lab Talk,* August, 1985, Vol. 29, no. 4, (Journal of the Science Teachers' Association of Victoria, Australia). Mr O'Reilly's article referred to *Science on a Shoestring* by H. Strongin, (Addison Wesley, 1976).

Using the OHP to show magnetic field patterns

G. A. Murphy, Our Lady's RC School, Oldham

The OHP can be a very useful tool for teaching about iron filing patterns and magnetic fields to middle secondary school pupils. Iron filing patterns can be prepared before the lesson on transparencies and preserved using the following technique:

1. Take an OHP transparency with cardboard mount.
2. Place the chosen magnet/magnets under the transparency and support the transparency using a shallow tray (see Figure). The transparency should be clean and level for best results.

3. Lightly sprinkle iron filings over the transparency—tapping gently in the usual manner until a good field pattern has formed.
4. Without moving the magnet, the field pattern can now be fixed on to the transparency by gently spraying with hair spray. Hold the aerosol can about 9 inches from the transparency and cover all the transparency with even strokes.
5. Allow a few minutes for the field pattern to dry.

If successful you will now have a permanent field pattern sealed on to the transparency. This can be covered with tissue paper and is then quite durable.

The OHP can be used to demonstrate the art of obtaining good field patterns, an art which many pupils find difficult.

A swinging wire

Geoff Auty

Many teachers of physics will have tried the kicking wire experiment [1] involving a suspended wire passing between the poles of a horseshoe magnet, and dipping into a trough of mercury which gave electrical continuity whilst allowing the wire the opportunity to swing in any direction.

Open dishes of mercury are not now recommended for use in schools, but I can offer an effective alternative.

The method I have found successful is to use a length of 'Extraflex' (eg RS Components 356167) or a number of such pieces joined with 4 mm stacking plugs (eg, RS Components 444539) as many school labs already use these for their stock of connecting leads. Pieces of Magnadur from a Westminster electromagnetism kit can provide the magnetic field, although other horseshoe magnets or pairs of strong bar magnets may be substituted. In fact to keep the region of the field small, I found it better to stack 2 pieces of Magnadur at each side of the iron yoke to provide the magnetic poles. The vertical wire is kept reasonably tight by tying a weight onto it. 20 to 50 g is suitable and a current of 5 A then provides a swing which is easy to observe.

The lower end of the suspended wire has to be laid loosely on the bench in such a way that the wire is free to swing in any direction.

Depending on the model used, internal resistance of the lab-pack may be sufficient to limit the current, but if teachers feel it to be a bad habit to short-circuit the power supply effectively using only the regular connecting leads, then an external resistor of 1 Ω to 2 Ω (depending on the supply pd) capable of carrying 5 A for a few seconds, could be included.

Although intended to replace a traditional demonstration, the use of standard laboratory equipment enables this method to be used as a

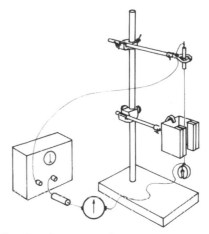

Figure 1   Arrangement of apparatus

class experiment if preferred, and is a very effective introduction to electromagnetic forces and the derivation of Fleming's left-hand rule.

It is a matter of preference whether to prejudge the result by telling the class which side to have the north pole, and whether to have the current flowing up or down. I leave it to chance and then ask why different class groups get different results.

One word of warning: the wire has to be suspended centrally between the magnetic poles with some accuracy. Otherwise due to the non-uniformity of the magnetic field, the wire will initially swing across the magnetic field (as the textbooks predict) but when out of the strongest region of the field, it will then fall towards the nearer pole.

Alas it is not quite as spectacular as the perpetual dancing if the 'kicking wire' with it resulting sparks, vapours, ions and consequent hidden dangers.

REFERENCE

1 eg Abbott, AF, *Ordinary Level Physics* (3rd edn), (Heinemann, 1980), p 443.

Geoff Auty, New College, Pontefract, West Yorks

The whirling neon lamp

R. D. Harris, Ardingly College, Sussex

I do not claim originality for this demonstration, but it is one that I have used many times over the years, and it always creates interest; in case it is not generally known, I venture to draw it to the attention of readers.

The experiment involves the use of a 5 W Osram 'beehive' neon night-light; this particular form of neon lamp does not appear to be currently available, but I expect that many school laboratories will still have one or two on their shelves; other types of neon lamp could doubtless be used, but the demonstration is particularly spectacular with the beehive model.

The lamp is fitted into a pendant holder, and connected to the mains via a long, flexible lead; the lead must be securely attached to the holder. The room is darkened, and the lamp is rotated in a vertical circle of any convenient radius (e.g. 0.7 m); I normally stand on the demonstration bench to do this, making sure that there are no obstructions in the

way! Spectators will see an effect somewhat like a catherine wheel. As the 'beehive' coil and the central disc alternately become the anode, they light up with the characteristic pink glow; these bursts of illumination can be seen spaced out over the circle at equal intervals, separated by patches of darkness which correspond to the time intervals when the tube is below the firing voltage. The faster the lamp is rotated, the wider the spacing of the patches of illumination. It is instructive to repeat the demonstration with the lamp connected to a 240 V d.c. supply, when a continuous circle of light is produced.

connected to a 240 V d.c. supply, when a continuous circle of light is produced.

I would normally precede the demonstration with a 'static' display of the lamp connected across a variable voltage d.c. supply; as the voltage across the lamp is gradually increased from zero, one or other of the electrodes suddenly lights up as the critical firing voltage is reached; I would then reverse the connections to show the other electrode lit up. Finally I would connect the lamp across the a.c. mains, showing both electrodes lit up apparently simultaneously but actually alternately, as the main demonstration shows so convincingly.

The photograph was taken by one of my sixth form pupils, Nicholas Adams. He used Kodak Tri-X Pan film (400 ASA), with an exposure of 0.5 second, at an aperture of f/4.5.

SAFETY NOTE

It is strongly recommended that the bulb is surrounded by an inspection-lamp cage, or similar device, in case of any untoward impact during this demonstration.

Teaching domestic electricity to the least able

J. Gillin and R. E. L. Waddling, Falmouth School

At present our least able fourth and fifth year students follow a course of Applied Science (SWExB — Mode 2) with limited GCSE grades C–G. The coursework is based largely on the *Science at Work* booklets [1]. For the weakest students (grade F and G), the section of the *Domestic Electricity* booklet that deals with domestic wiring circuits can be rather daunting. While it is of value to construct working lighting circuits using household switches, the complexity of the assembly operation can often be beyond these pupils, particularly when it involves wiring on two sides of a peg board.

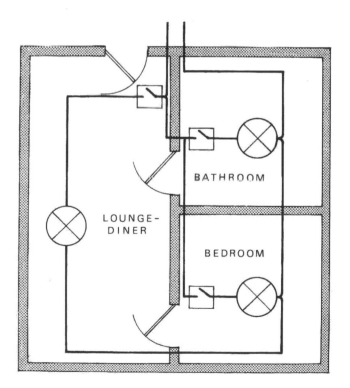

LOUNGE-DINER

BATHROOM

BEDROOM

We have found that by using simpler pieces of equipment, students are able to gain a better understanding of the very basic principles involved. These we see as being able to appreciate that:

1. Electricity flows around a house in a circuit.
2. In an electrical circuit all components are joined to both terminals of the power supply.
3. Bulbs should be wired in parallel.
4. A fault in one bulb (in parallel) does not affect the others.
5. Extending a circuit to include more bulbs (in parallel) does not reduce the brightness of the original bulbs.

Students are provided with a piece of formica-coated chip board (approx 30 cm square) on which has been drawn a house plan using an indelible OHP pen (see Figure 1). They are also given a self-sealing plastic bag containing a small electrical screwdriver, wire strippers, four MES bulb holders with 6 V bulbs, four 1 amp push switches, and a lump of Blu-tack. Each house plan and all components are numbered to ease the organization.

Students, usually working in pairs, construct the lighting circuit using a communal supply of bell wire and the Blu-tack to position the wire in place. Connections are made by joining more than one wire into appropriate terminals. The low voltage laboratory supply is used as the power source.

We have found that students enjoy this activity. The task is presented in a more meaningful way than by just using traditional circuit boards and also it is immediately obvious to the student what has to be done. This activity can be used to assess practical and problem-solving skills and will take most students a double lesson to complete.

There are two possible extension activities. The first of these is a problem-solving activity, in which students are asked to divide the lounge into two rooms and to wire in the fourth lamp and switch. Students who successfully complete this use the reverse side of the board to design and wire their own house. The plan can be drawn on with water-soluble OHP pens and washed off at the end of the lesson. With guidance most students should be able to wire in the fourth lamp and switch during a second lesson. Faster groups will be able to design their own houses and lighting circuit.

This equipment has also been used successfully with single certificate Science (16+) candidates and could be used with sixth-form CPVE students.

SAFETY NOTE

Students should, of course, be reminded that the switch in the bathroom must only be operated by a remote control, usually a length of cord.

REFERENCE

1. Allen, R., *Domestic electricity, Science at Work* (Longman, 1979) p.11.

Eddy current damping - the long and the short of it

Iain MacInnes

The Westminster Electromagnetic Kit includes a number of simple, informative experiments on eddy current damping [1, 2]. Specifically, the damped oscillations of a spinning aluminium ring are examined.

An instructive variation is to observe the oscillations of the ring when it is used as the bob of a simple bifilar pendulum as illustrated in Figure 1. The procedure is to release the ring from above a fixed mark and observe the damping due to air resistance only Figure 1(a). Pupils can be asked to count the number of oscillations required for a noticeable decrease in the amplitude of the swing.

Next, they are asked to predict whether the oscillations will be damped more heavily when oscillating through the length of the magnadur magnets Figure 1(b) or through the breadth as in Figure 1(c). Many pupils will surprised to discover that the damping is greater in the shorter field of Figure 1(c). The reason is that after the amplitude has decayed to the stage when the ring is oscillating completely within the length of the magnet, there is no change of flux and damping is again due to air resistance only. The experiment emphasizes the important point that there will be no induced emf and therefore no eddy currents when the flux remains constant. A useful question to ask is 'What is the ratio of length of magnet to diameter of the ring which will provide the greatest damping?'

REFERENCES

1 Westminster Kit Instructional Booklet. Available from main suppliers of laboratory equipment.
2 Jardine, J, *Nat Phil Book 4*, (Heinemann Education Books, 1974), p 209.

Iain MacInnes, Head of Physics, Jordanhill College of Education, Glasgow

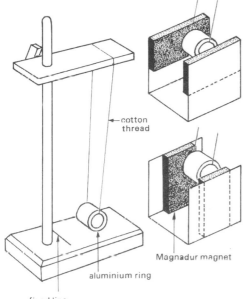

cotton thread

Magnadur magnet

aluminium ring

fixed line

Figure 1(a)

Figure 1(b)
Figure 1(c)

Investigating the discharge of a capacitor with a flashing neon - a footnote

DL Thompson

Physics teachers will be familiar with the experiment in which a neon bulb is used to investigate the discharge of a capacitor (see for example *Advanced Level Practical Physics* [1]).

I recently came across an interesting historical use of this method. The elegance and ingenuity of the experiment is striking. The use is described in [2].

The neon lamp is used to test a capacitor for leakage in the following way. The circuit shown in the figure is set up.

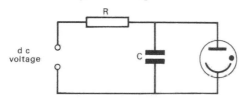

The neon lamp will flash at a rate determined by the time constant CR and its own characteristics. A 'suspect' capacitor of sufficient working voltage is now placed in parallel with 'R'. If the rate of flashing of the neon is not appreciably altered then the insulation of the capacitor is good. A measurable alteration in the flash rate enables the leakage resistance to be estimated.

In the early 1930s the home radio constructor may well not have possessed a measuring instrument such as a milliammeter and a method such as this was really ingenious.

REFERENCES

1 Nelkon, M and JM Ogborn, *Advanced Level Practical Physics*, 4th edn, (Heinemann, 1978) p 176.
2 Scott-Taggart, J, *The Book of Practical Radio* (London: The Amalgamated Press Ltd, 1934).

DL Thompson, Faculty of Education, Huddersfield Polytechnic, Holly Bank Road, Huddersfield HD3 3BP

Instructions for making a miniature electric motor

Rosemary Bates

EQUIPMENT

Two anisotropic ceramic ferrite block magnets. (These are also called magnadur magnets – they have their poles on the large faces.)
Two 5 cm safety pins
A reel of 0.56 mm diameter enamelled copper wire (24 swg)
A rod or tube approximately 1 cm in diameter for winding the coil
  (a test tube or an AA size cell might serve)
One C size 1.5 V cell
A short thick rubber band (eg Velos No 70 - 38 mm long 9.5 mm wide)
Scissors
Metre rule
Glasspaper
Masking tape
Plasticine or polystyrene cup

METHOD

Cut off a metre length or wire
Scrape off 5 cm of enamel insulation from each end of the wire, using glasspaper.
Wind the wire around the rod, test tube or other chosen former, leaving the scraped ends protruding by 5 or 6 cms.
(Each successive turn of wire touches or overlaps previous ones, and should not be spread out along the surface of the former)

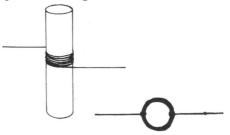

Slide wire coil off the former, holding it together with the fingers.
Loop each free end of wire once through the coil to stop it unwinding.
Lay the C-sized cell on its side, on a table.
Turn a safety pin up-side-down.
Rest the pin head on the table against one end of the cell.
The pin must touch the terminal, and point vertically up into the air.
Tape the pin to the cell.
Attach a second safety pin to the other end of the cell in the same way.

(Make sure the hinges of the pins line up.)

Ensure a good connection of pins to terminals with rubber band.

Stop the cell from rolling, by wedging it with plasticine on either side, or by taping it directly onto the table top, or inverted polystyrene cup.

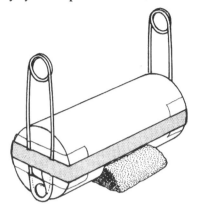

Insert the free ends of the coil of wire into the hinges of the pins.

Make sure the free ends of wire are horizontal, and the pins are level.

Flick the coil of wire by means of your fingers to see if it spins freely.

Adjust if necessary.

Rest a magnet on top of the cell, as shown in the diagram.

Spin the coil. (Several attempts may be necessary to get it going.)

When you have got it to work, try stacking two magnets on top of the cell.

Note the effect on the rotating coil.

Other positions and arrangements of the magnets can be tried out.

copper wire scraped bare with glasspaper

insulated copper wire

5 cm safety pin

elastic band

masking tape

Plasticine

anisotropic ceramic magnet

1.5 V cell MN1400 HP11 (size C)

REMINDERS

1 Do not leave the coil resting on the pins without operating the motor, as this will soon run the cell down.

2 When several cells have been used by a class, they should not afterwards be used together as a battery in any other piece of equipment, game or toy.

This is because the cells may have run down unevenly. It has been known for explosions to occur when long-life alkaline cells with a high electrical charge are mixed with nearly discharged cells. A reverse charge current may be set up which results in a build up of gas pressure in the sealed casing which then bursts.

ACKNOWLEDGEMENTS

My thanks to H Strongin and GD Giffould.

REFERENCES

1 Strongin, Herb, *Science on a Shoestring*, (Addison Wesley Publishing, 1985).

2 Giffould, GD, 'An electric motor model that really works!' *SSR*, 1987, **69** (247), 332-4.

3 Health and Safety at Work Bulletin, September 1987, Inner London Education Authority.

R Bates, Tower Hamlets ILEA

Electroscope surprises

Colin Siddons

THE CAN ELECTROSCOPE [1, 2]

The can electroscope has some advantages over the traditional Gold Leaf Electroscope. It costs nothing: it is easily made, it can be seen by a whole class. Unlike the GLE it shows surprising leakage effects.

A can rests on an insulator, eg a clean dry plastic tub. Two arms hang from its sides. They must be suspended so that they are free to

swing outwards: this can be done by hanging them from a pin whose pointed end is held in a block of Blu-tack.

Suitable *metal* arms can be cut from 'silver paper' eg the wrapping paper from bars of chocolate. *Paper* arms can be cut from tissue paper, however, they do not show leakage effects as well as metal arms do.

To show the can electroscope as a charge detector half-fill a plastic container with cake beads (or grains of rice). Shake vigorously. Pour the beads into the can. The arms rise slowly. Children find this very amusing (see figure).

To charge the can *positively* a glass vessel of suitable shape and size is needed. By Franklin's historic choice glass when rubbed with silk becomes positively charged; but not all glasses charge - and not all silks are really silk. A search amongst glasses and silks will eventually unearth a charging glass-silk pair. Rub the glass vessel vigorously and lower it into the can. The raising of the arms will signal success.

To charge the can *negatively* fold a polythene sheet of suitable thickness into a cylindrical shape. Use glue to keep it in this shape. Rub the cylinder with flannel or wool. If you have a dry skin you can charge by hand. (Don't ask pupils with moist hands to help you with electrostatic experiments.)

First surprise

This is for junior forms. Lower a negatively charged cylinder into a can electroscope fitted with thin metal arms. The arms deflect.

Ask a pupil to discharge the can by touching it for a moment. The arms drop. Remove the cylinder. To the pupil's surprise - if not the teacher's - the arms rise again.

Second surprise

Lower a well-charged negative cylinder into the can but this time note the size of the deflection. Touch the can, remove the cylinder to get the second deflection. Note its size. Without measuring it is obvious that the second deflection is bigger than the first. This is surprising for charged bodies in air gradually lose their charge, so the second would be expected to be smaller than the first.

Third surprise

If instead of using a negative cylinder a positively charged glass beaker is used, this time the second deflection is less than the first. This is only a surprise because it is the opposite of the previous case.

Fourth surprise

Again lower a well-charged negative cylinder into a can. This time, however, slowly remove the cylinder WITHOUT TOUCHING the can. At first - no surprise here - the arms drop but as soon as they reach the bottom they rise again.

I showed this experiment at the Lancaster Meeting of the ASE. Even veteran physics teachers were surprised.

EXPLANATIONS

The *first surprise* neatly demonstrates induction. When the negative cylinder is lowered into the can it induces a positive charge on the inside of the can.

Since the can is standing on an insulator this positive charge must be balanced by a negative charge on the outside of the can. Touching the can allows the negative charge to flow to earth, the positive charge still being held in place by the charged cylinder. Removing the cylinder allows the negative charge to flow to the outside.

The *fourth* surprise is due to the leakage of charge from the sharp edges of the projecting metal arms. As soon as the arms deflect they lose some negative charge into the surrounding air. There will now be more positive charge inside the can than negative charge outside. When the cylinder is removed this greater positive charge is set free and drowns the smaller negative charge. The arms deflect with positive charges.

To confirm this explanation use a cylinder that is taller than the can. Charge it. lower it slowly into the can. The arms deflect at first but soon reach a *maximum* deflection. More positive charge is being induced inside the can but the negative charge on the outside has reached its limit. It is holding as much charge as it can: further lowering merely results in more leakage into the surrounding air.

The *second* surprise shows that there is a difference in leakage between positively and negatively charged bodies. We have seen that due to leakage the arms have a maximum angle of deflection. The maximum angle for positive charges is greater than the maximum angle for negative charges.

The angle of deflection of the arms when a negative cylinder is put in the can is a measure of the negative charge remaining on the outside *after leakage*.

When the can is touched and the cylinder removed the full positive charge flows to the outside. The fact that the deflection now is greater shows that the leakage from this posi-

Oh no, not more potential dividers!

Adrian Watt

This article outlines an experiment where the voltage from the mid point of a series of potential dividers is connected to the, previously calibrated, frequency modulation input of a signal generator. By matching the input voltages with those needed to produce the notes in a musical octave a simple 'electronic organ' can be constructed. Not only does this approach give practice in calculating values for the resistors in two resistor networks; it also introduces the idea of 'preferred' resistors.

THE BACKGROUND

The concept of the potential divider is central to the understanding of many ideas in both physics and electronics. Even those pupils who have been brought up on a diet of 'input sensor modules' have difficulty when they want to know more and are introduced to the potential divider. One common way around some of these conceptual problems is by using analogies to try to simplify the situation. However, some of these are more successful than others. Frequently the best way to consolidate the concept of the potential divider is to calculate an expected output value and then investigate the arrangement experimentally. But even this approach can become tedious with many seemingly meaningless calculations. This experiment involves many of the same calculations but these are now seen by the pupils as having a definite purpose.

CALIBRATION

Many of the signal generators which are commonly available in schools have an internal frequency modulation facility. This is primarily intended to allow an external voltage ramp to sweep the output frequency through a wide range of values. If a variable low voltage power supply is connected to the modulation input the output frequency can be changed manually. If a digital frequency meter is connected to the output and digital voltmeter across the input, the modulating voltages needed to achieve the frequencies shown in Figure 1 can be found. As these frequencies correspond to the notes in the octave above 'middle C' the system can now be regarded as calibrated.

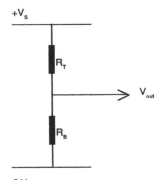

Note	Frequency/Hz
C	256
D	288
E	320
F	341
G	384
A	426
B	480
C	512

Figure 1

Figure 2

THE CALCULATIONS

For the simple 2 resistor network shown in Figure 2 the voltage at the mid-point is found using the equation

$$V_{out} = \left[\frac{R_B}{R_T + R_B} \right] V_s$$

If the system is to produce the eight notes in the octave, eight of these potential divider arrangements need to be constructed and one suitable arrangement is shown in Figure 3. The calculations can be simplified if the lower resistor in each arrangement is kept the same. (100 kΩ should give suitable versatility and accuracy) Once the values for the lower resistance and the supply voltage have been decided, the values of the other resistors (R_1 to R_8) can be calculated. These will be the values needed to set points A to H at the potentials determined during calibration.

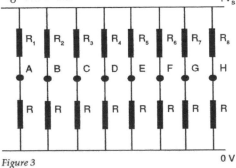

Figure 3

There are many possible methods for assembling the resistors and playing the different notes. The most successful in our set up involved assembling the resistors on a 'bimboard' type circuit board and using the probe lead from a multimeter as the 'plectrum'. A loudspeaker must also be connected to the output of the signal generator.

POSSIBLE PROBLEMS

The frequency modulation input of some signal generators is resistive. This means that when the potential divider is connected the input voltage is altered. This problem can be overcome by 'buffering' the frequency modulation input using the simple op-amp follower type circuit shown in Figure 4.

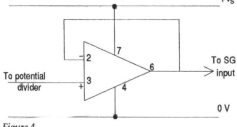

Figure 4

CONCLUSION

The importance of potential dividers justifies a significant quantity of time being devoted to them. The approach outlined in this article has proved successful particularly with the more able pupils and the extra time involved was considered well worthwhile. The most fun was derived when one successfully completed system was used in a game of 'Name that tune'.

A Watt, Merchiston Castle School, Colinton, Edinburgh EH13 0PU.

To start with, electrical measurements on my jug kettle, rated at 2 kW, show that the resistance (Avometer) is 28 Ω, while the current drawn is initially 8.4 A, when the kettle contains water at 20 °C. These values hardly change as the kettle boils, and give power ratings of 2.1 kW and 2.0 kW respectively.

For thermodynamic measurements, the kettle is made of thin white plastic, which is insulating and of low emissivity, and also of low heat capacity. It is conveniently calibrated in litres, so that it is easy to use the power rating to calculate the specific heat capacity of water. A temperature rise of 45 °C in 100 seconds for 1 kg of water gives a value of 4500 J/kg K which is acceptably close and also acceptably high. Domestic tranquillity prevented my experimenting with other liquids.

It is when you bring it into partnership with the electronic top pan balance that the jug kettle excels. The essential characteristics of the kettle here are its lightness with the top removed (which also prevents it from switching off), and its ability to boil small quantities of water. These allow you to observe boiling on a typical balance which has a maximum load of 1.5 kg. Zeroing the balance when boiling starts allows readings of mass loss to be taken periodically for one or two minutes. My kettle loses 0.87 g/s, which gives the specific latent heat of vaporization of water as 2.3 MJ/kg.

For about double the price of a standard low voltage immersion heater you can therefore get a purpose built calorimeter with integral immersion heater, designed to illustrate a number of principles of heat transfer, and which doesn't need a separate power supply. And you can make a cup of coffee afterwards...

J Miller, Tasker Milward School, Haverfordwest SA61 1EQ

Jug kettles in the science lab

John Miller

God invented the nasturtium and the onion ex pressedly for biology teachers, I've been told Whether He had a direct hand in jug kettles I am not sure, but they certainly seem tailor made fo some aspects of physics.

Demonstrations with LEDs

John Bolton and Jim Ashworth

It is worth keeping an eye on the latest electronics catalogues for new products to help make the teaching of physics and electronics more interesting. The following notes may be helpful to bring teachers attention to some novel LEDs (light-emitting diodes) and ways of using them to demonstrate various fundamentals of electricity and electronics.

1 BIDIRECTIONAL RED/GREEN LEDS

These are really two LEDs in a single package indistinguishable in appearance from a standard LED. Red light is given out when connected to a battery one way and a green light the other. It is therefore a good indicator of the directional nature of electric current. If given to pupils – made up in a module with a protective 200 Ω series resistor they enjoy testing this for themselves. An additional demonstration is to connect the module to an ac source which gives an orange-yellow output, explained by the mixing of two primary colours over the red to green section of the visible spectrum.

2 UNIDIRECTIONAL RED/GREEN LEDS

These are also two LEDs in a single package but with both red and green acting in the same direction. There are therefore two anodes and a common cathode. Connected in a circuit as shown in Figure 1 the currents in the red or green section can be varied to give any colour of the spectrum between red and green. This can be

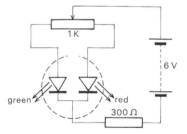

Figure 1   Unidirectional R/G LED in a voltage-divider circuit

used to illustrate the functioning of a voltage divider.

3 HIGH-INTENSITY LEDS

These high-intensity (ultrabright or hyperbright in the catalogues) will show a discernible red light at extremely low currents of one microamp or less. This makes them very useful to illustrate the following:

(a) Charge storage in a capacitor. After charging a 2000 microfarad capacitor from a 6 V battery it can be discharged through the LED (with a 200 Ω series resistor) to give a visible exponentially fading light for about three minutes. Of course all the polarities must be correct!

(b) The measurement of very low currents. The largest resistor in series with the LED when connected to a 6 V battery to give a clearly visible light is more than one megohm. This can be compared with the least current to give a visible light from an incandescent lamp of about 100 milliamps. Pupils can work out for themselves the power of these lamps - and the differences between 'electronics' and 'electricity' in terminology (microwatts, milliwatts and watts).

(c) The same LED and resistor module can be connected to two copper/zinc plate cells with an electrolyte of some kind to give a bright light for a considerable time. This is because it requires such a small current that polarization is slow. (Another relatively new component which will work in the same circuit is a low-current piezo-buzzer, again using only a few milliamps. The sound is a bit tiresome after the first 20 minutes!)

4 FLASHING LEDS

Red or green flashing LEDs look no different from the standard 5 mm types but they have integrated circuits incorporated in them which cause them to flash at about 0.5 Hz. The IC is powerful enough to 'flash' two more standard LEDs in series with a voltage drop of about 1.2 V for each LED. No series resistors are required. These LEDs illustrate the miniaturization now possible with ICs, especially if the flasher is compared with a standard LED in an astable circuit made from transistors.

Children seem fascinated by flashing lights and they should be aware of this 'simple' way of introducing them into other circuits.

5 AN LED RECTIFIER

Previous articles in *SSR* by Ashworth and Thompson [1] and Green [2] have described how LEDs can be used to illustrate the working of a full-wave bridge rectifier, with red LEDs showing one half-cycle and green LEDs the other. A yellow LED is connected to the output. This arrangement is excellent for showing not only how a rectifier works but also to show visually the waveform of the input. When connected to a frequency and waveform generator set to a low frequency, the difference between sine-wave, square-wave and sawtooth-wave inputs can be clearly distinguished by the way the diodes switch on and off.

6 LASER DIODES

Infra-red laser diodes have been available for some time and Bishop [3] described their use with a suitable drive circuit for interference and diffraction experiments as well as for line-of-sight communications. These lasers work at wavelengths of about 900 nm in the near infra-red, but the use of such invisible laser light that could be dangerous would be unacceptable in schools. Recently visible red light (670 nm) laser diodes having become available at a reasonable price – these are the lasers used in supermarket bar-code scanners. 630 nm laser diodes are also available, producing light of similar wavelengths to the He-Ne gas laser (633 nm) without the need for a high-voltage power supply. It will not be long before these will be cheap enough to be used in schools. (A 1 mW low voltage visible laser diode, packaged with a pulsed power supply and focusing optics in a small 75 × 15 mm unit is now in the RS catalogue but is still quite expensive.) The laser diode itself costs about £15. Of course strict safety precautions must be observed with ANY laser.

7 MOUNTING THE DIODES (AND OTHER COMPONENTS)

LEDs like other components and modules need mounting in some way for them to be connected together or to other commercially produced

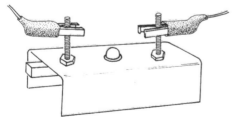

Figure 2   Method of mounting the LEDs and other components

modules. The method we have found both cheap and convenient is based on cut 8 cm sections of square-section polythene drainpipe available in black or white or brown. A standard 3 m length is enough for about 68 mounts. The sections are cut longitudinally to give U-shaped cross-sectioned units with a top area of 8 × 6 cm which is exactly half a 5 × 3 in filecard which can be used as a template to standardize the layouts of the components and connections. A typical layout is shown in Figure 2. Wiring connections are made using M3 nuts and 12 mm bolts and M3 wire tags on the mounts and flexible wire with plastic-coated mini-crocodile clips. These are available very cheaply from most component suppliers. Using this system, three or more connections can be made to each terminal if required by the circuit.

For a class display or demonstration the modules can be supported inclined towards the class

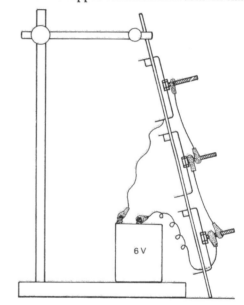

Figure 3   Method for class display of the modules

on a plastic-coated wire frame (an old refrigerator shelf). This is illustrated in Figure 3.

REFERENCES

1 Ashworth, K and DL Thompson, 'A demonstration of bridge rectification', *SSR*, 1986, **67**(240), 589.
2 Green, CV, 'Looking at ac and dc with light-emitting diodes', *SSR*, 1986, **68**(243), 297.
3 Bishop, O, *Electronic Science Projects*, ch 1 A simple infra-red laser (Bernard Babani (Publishing) Ltd, 1983) p 104.

J Bolton and J Ashworth, St George's School, Harpenden, Herts

A method for converting British overhead projectors to wide angle

A. J. Herring, Sunderland College of Education

The principle is simple. You remove the front lens and place it underneath the bottom lens.

PROCEDURE

Removal of front lens

(a) Tilt the adjustable part of the head right back.
(b) Insert the fingers into the slot, loosen the screw at the top centre and allow the spire nut and clip to fall out.
(c) Withdraw the fingers, bringing the lens with them.
(d) Reassemble the screw, clip and spire nut in place, adding half a terry clip, size 80/0.
(e) Slide a piece of picture glass, $3\frac{1}{2}$ inches in diameter, into the place formerly occupied by the lens, and tighten the screw.

Manufacture of lens mount

Cut 2 pieces of 16 or 18 s.w.g. aluminium 4 in × $3\frac{3}{4}$ in, and cut a $3\frac{1}{4}$-in diameter hole in the middle of them. Assemble as shown in Fig. 1.

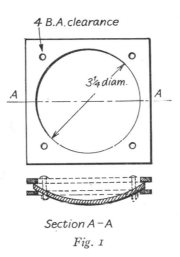

Section A–A

Fig. 1

Manufacture of carrier for lens mount

Cut 2 pieces of 16 or 18 s.w.g. aluminium $4\frac{1}{2}$ in × $1\frac{1}{4}$ in, and bend as in Fig. 2. One must be a mirror image of the other.

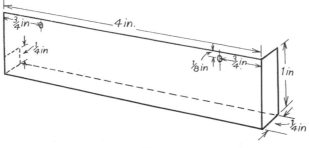

Fig. 2

They are fixed to the bottom half of the projector head with either pop rivets, self-tapping screws, or nuts and bolts (difficult without dismantling the whole head). The lens mount now slides in under the bottom lens.

If it is desired to be able to return the projector to normal optics occasionally, a

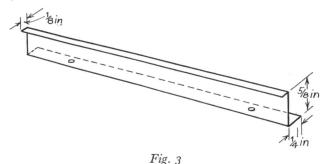

Fig. 3

further piece of aluminium 4 in × 1 in bent as in Fig. 3 is needed. It is fitted to the bottom edge of the top half of the head by the method previously used.

MATERIALS USED

16 or 18 s.w.g. aluminium:
  4 in × 3¾ in (2 pieces)
  4½ in × 1¼ in (2 pieces)
  4 in × 1 in (1 piece)
  4 × 4 B.A. nuts, bolts and washers ¾ in long
  6 pop rivets, or whatever is to be used
  1 terry clip, size 80/0

Storage of small fragile items

P. Gee, Dr Challoner's Grammar School, Amersham

Whilst I was on a course at Worcester College of Education, I was shown, by Mr E. J. Wenham, the following method of storage. We have subsequently used it very successfully and, as I have not seen the idea elsewhere, I mention it here.

Acquire—from the local market—a sheet of foam plastic (as used for cushions, etc.); we find 1 to 1½ inches thick the most useful. Cut a piece to fit in your storage trays (we use the grey moulded plastic ones but cardboard ones will do). With a scalpel (from biology) cut holes or slits in the foam so that the items will fit in comfortably. Now stick in the foam with Uhu.

We find that a single cut is suitable for the conventional lenses, but the cylindrical ones require holes to be made. Other items may require large pieces to be cut out. A single 18 in by 11 in tray will hold all 80 lenses or 32 cylindrical ones. We put blobs of paint to identify different types of lens and put similar blobs, together with Dymo-tape labels inside the tray.

The method is ideal for lenses, mirrors, magnets, scissors, wire strippers, Westminster kits, etc. Missing items are immediately noted and the lads take care in putting items back in the correct niche.

Clamping devices

W. K. Mace, King Edward VII School, Sheffield

Two simple gadgets have proved their usefulness and versatility over several years, and may be of interest. The first (Figure 1) was made to simplify the support of simple pendulums. The block is about 3 cm × 3 cm × 1 cm, and the hardboard rides loose on the screw with about 2 mm play between it and the block. A standard 3-jaw clamp holds the two tightly together and provides a precise pivot point for the emerging string or cotton.

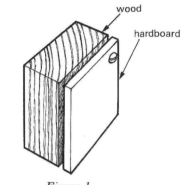

Figure 1

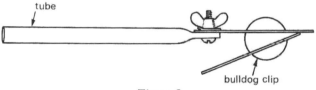

Figure 2

The second (Figure 2) may commend itself to anyone who has tried quickly to set up a periscope on a laboratory stand! It consists of a length of alloy tube pinched at one end, a wing nut, bolt and washer, and a large bulldog clip which has had one of its jaws ground down so that any flat object is held positively in the plane of the other jaw. A mirror may thus be clamped in a bosshead so that it can be rotated reliably about a line in its own plane without any sideways displacement. We keep a couple in each lab for holding mirrors, glass plates, ground glass screens, slides for diffraction, etc.

Storing leads

Peter Griffiths and John Stephens, Ladymead School, Taunton

Being dissatisfied with the normal peg board system for storing 4 mm leads we have developed the following idea. For lower school experiments it is unusual to need more than five leads. Therefore if the leads could be stored in sets of five they would be easy to issue and to check on return.

The storage method involved the use of a strip of plastic book binding material with a

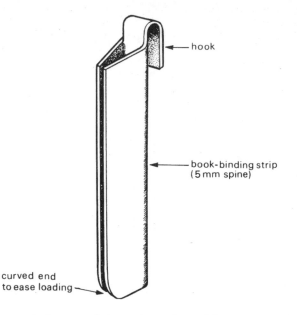

hook

book-binding strip
(5 mm spine)

curved end
to ease loading

5 mm spine. This was cut at one end, heated gently and bent over to form a hook. The other end of the strip was rounded off to facilitate the loading of the leads.

To put a lead into the holder the lead was placed against the curved end and by running the thumb along the strip the lead is pushed in. This is repeated until all leads are in place, and the holder is then hung on a rack.

We have found it much easier to check leads for faulty plugs and have found that the method reduces plug loss and lead tangle almost completely.

Keepers on the eclipse major

M. D. Ellse, St Paul's Girls' School, London

I have suffered various minor injuries and inconvenience trying to remove the keepers from our Eclipse Major magnet. Pliers are not always to hand and even sliding the keepers off is impossible for many pupils. Recently I drilled a hole through each keeper and screwed them to a piece of wood. That seems to have solved the problem.

Pivots for science apparatus

John Mattocks, Bristol Polytechnic

One of the problems met in making science apparatus for use in schools is the provision of a cheap, robust, low-friction pivot. In relevant literature the female part of a snap fastener is often suggested. I have never found this to be entirely satisfactory. An old ball-point pen and a combustion tube are much more successful and cost little.

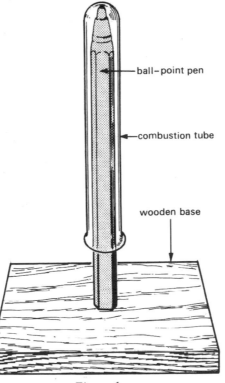

Figure 1

A combustion tube of internal diameter a little wider than the pen is needed. The pen is sawn off about 2 cm longer than the tube and is glued into a hole in the centre of a wooden base. As an alternative to the tube, a length of glass tubing which has had one end sealed by rotating it in a bunsen flame is equally effective.

Suggestions for the kind of apparatus for which the pivot is useful include a compass Figure 2(a), a fan for convection currents Figure 2(b) and a device to show electrostatic attraction and repulsion Figure 2(c).

For the compass, two discs of cardboard are cut out and old razor blades which have been magnetized are glued between them. A hole is made in the centre of the discs with a cork-borer of diameter appropriate to achieve a push fit on to the combustion tube. Obviously it is necessary to align the razor blades so that they are parallel and have their polarities in one direction. The magnetic axis can be marked on the upper disc with a felt pen.

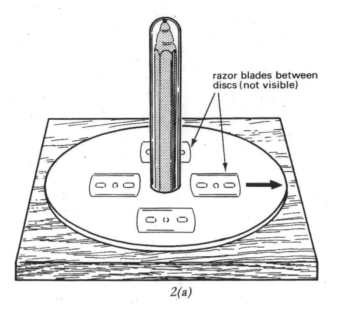

razor blades between discs (not visible)

2(a)

In the electrostatic apparatus a plastic knife is fixed to the tip of the combustion tube with a dab of glue, or with a loop of sellotape. Plastic spoons will work, but are not quite as long as the knives.

For the convection fan, instead of using a wooden base, the pen can be held in a retort stand and so a low bunsen flame, or a candle can be used as a source of gentle heat below. The friction of the pivot is so low that rotation can be achieved even with the heat from a 60 W light bulb.

No doubt readers can device other uses for this practical pivot.

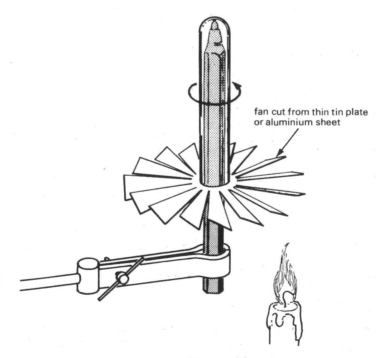

fan cut from thin tin plate or aluminium sheet

Figure 2(b)

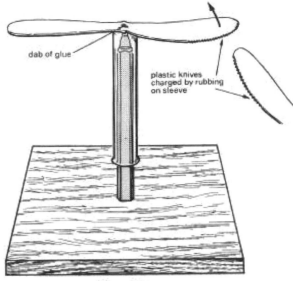

Figure 2(c)

A useful gadget for small soldering jobs

PJ Sapwell

I have devised a simple gadget made out of clothes pegs as an aid to small soldering jobs which one often has to do when making Physics Electric/Electronic equipment. It is obviously similar in principle to the commercially available 'Helping Hands', but being made of wood, my gadget does not conduct away the heat of the soldering iron.

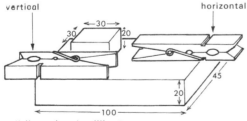

all dimensions in millimetres

I use it regularly when soldering wires on to LEDs and 2 mm and 4 mm plugs.

PJ Sapwell, Beeston Hall, West Runton, Cromer NR27 9NQ

Plastic bottles as aspirators

RB Marks

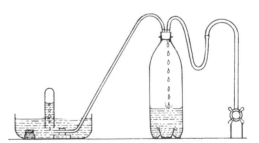

Figure 1

This method (Figure 1) of dispensing relatively insoluble gases offers a number of significant advantages.

MANAGEMENT OF GAS RESOURCES

1 Plastic soft drink bottles are well designed to act as a convenient store or stockpile of gas(es), being designed to hold pressurised gas for long periods while drinks are in the retail distribution network. In this respect, they are superior to other plastic containers which could be used.

Simple qualitative tests indicate that gases such as oxygen and carbon dioxide remain unaltered for at least a month and probably much longer, provided they are firmly stoppered.

Different gases can be indicated by using different coloured stoppers and rings (Figure 2).

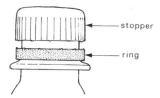

Figure 2

The author's choice was:

RED = OXYGEN;
BLUE = NITROGEN;
WHITE = CARBON DIOXIDE.

Since all common gases are colourless the choice is arbitrary and can be decided on availability.

If bottles of the same size are used, they stack well in a cupboard or on a shelf etc. With stoppers outwards a bottle of the required gas can be selected instantly.

The bottles are light and in normal use virtually unbreakable. Supplies can be obtained from domestic consumption, appeals to pupils and from plastic bottle banks which are common in many places.

2 Different sources of a gas or gases can be utilized to fill the bottles.

The author's experience is limited to pressurised cylinders of oxygen, nitrogen and carbon dioxide. Gases generated by chemical means can also be used, however, for example carbon dioxide from acid and marble chips.

3 If the system is used properly, systematic gas wastage (that is wastage or loss incurred as a result of methods used to deliver gas to the teaching area) is minimised or eliminated. The system gives best results when a supply of gas is required over a period of time.

Initially, the bottles are filled with water which is displaced as the gas is bubbled in. This should be the only time the bottles are deliberately filled with water.

As the gas is dispensed in the teaching area, the bottle becomes filled either wholly or partly with water. In either case the water and any remaining gas should be left and the bottle firmly stoppered before returning it to the preparation area. This procedure means that bottles never have to be refilled with water, and none of the unused gas is wasted (Figure 3).

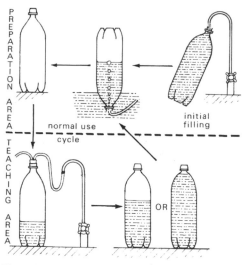

Figure 3

ADDITIONAL NOTE

For reasons of brevity, much detail has been

High-speed photography

J. K. Easlea, Portsmouth College of Education

A xenon stroboscope, such as the Dawe model 1214A, can be used to provide the illumination for single-shot photographs of fast-moving objects such as pellets from an air-rifle or of collapsing glass from a lamp struck by a steel ball.

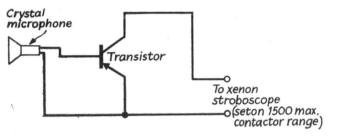

Fig. 1

One of the simplest methods employs an Acos crystal microphone connected to the base of a transistor (a Newmarket NKT 713 works well). The collector and emitter terminals of the transistor are joined to the 'contactor' socket terminals of the strobo-scope. Noise from the firing of the air-rifle is picked up by the microphone producing a voltage pulse which causes the transistor to conduct and 'switch on' the stroboscope momentarily. Good results were obtained using a Polaroid 800 camera, at aperture EV13 and shutter set to B, on 3000 ASA film. The microphone to rifle muzzle

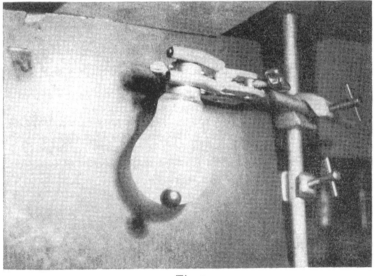

Fig. 2

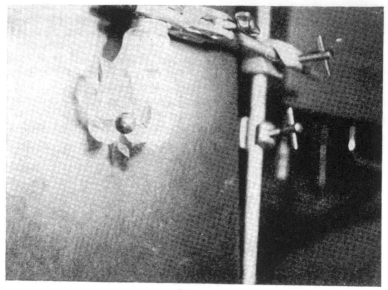

Fig. 3

separation needs to be at least 0·8 metres to provide sufficient time delay between noise and light flash for the lead pellet to escape from the barrel of the air-rifle so that it can be photographed.

The photographs above were taken with a 35-mm camera at full aperture using Ilford HP3 film.

Thanks are due to Mr L. R. Baker and Mr C. Moore for their willing assistance in trying out the technique and in taking photographs.

'Particles' of potassium permanganate

P. D. Arculus, Solihull School, Warwickshire

'Analoids' [1] of potassium permanganate are an ideal starting material for Fowles' classic dilution experiment [2] since they permit the usual calculation of the *number* of 'particles' to be extended to a determination of the *size* of the 'particles'.

Junior forms will be able to find, from the experiment, that a 0.5 g Analoid must be divisible into at least 10^7 parts, and, by measurement of its diameter and height, that its volume is about 10^{-1} cm^3. Thus the volume of each potassium permanganate 'particle' is 10^{-8} cm^3 and the linear dimension is 10^{-3} cm, approximately. There is no need to disclose that the potassium permanganate 'Analoid' contains a disintegrator or inert filler.

At a later stage, when the significance of Avogadro's number has been appreciated, it may be calculated that a 0.5 g Analoid must contain at least 10^{21} 'particles'. This, with the previous dimensional measurements, leads to a 'particle' volume and linear dimension of the order of 10^{-22} cm^3 and 10^{-8} cm respectively.

REFERENCES

1. 'Analoids' are compressed tablets of pure reagents of stated composition (some, like the potassium permanganate tablets, contain a disintegrator or an inert filler). They are obtainable from Ridsdale & Co. Ltd, Newham Hall, Newby, Middlesbrough, Teesside.
2. Fowles, G., *Lecture Experiments in Chemistry* (Bell, 1963), 15.

Physics of the 'Goofy Bird'

Alan Ward, Saint Mary's College, Cheltenham

The diagram below is of a restless 'Goofy Bird', which, as long as the beaker D is filled with water, keeps dipping to 'drink'. Glass bulbs A and B, connected by a tube C, comprise a sealed body containing a quantity of a very volatile liquid (ether would do, but is extremely inflammable). Head and beak are covered with felt-like material and this must be kept very wet. The body parts are pivoted to a static pair of plastic legs.

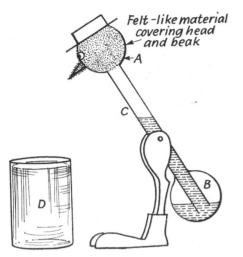

Felt-like material covering head and beak

Most readers have probably seen a Goofy Bird toy in action. But how does it work? Warmed by its surroundings, the volatile liquid evaporates and causes an increase of pressure inside B, which forces liquid mass to go up the tube and into the bottom of the head. Upward displacement of mass raises the centre of gravity to above and just in front of the pivoted 'axis'—thus making the bird bob down and dip its beak into the water. Immediately, an exchange of vapours between A and B is possible, and the liquid is able to run back to B.

With its centre of gravity below the axis again, the bird nods upright—and previous actions are repeated. Meanwhile, water evaporating off the wet head takes latent heat from A, keeping the space cool and causing some of the vapour inside it to condense—therefore stopping any build-up of pressure in A. Capillarity, via the regularly re-wetted beak, keeps the felt moist throughout.

A Goofy Bird can be bought for as little as 50 pence at a toy or joke shop. It can be used in school science as a starting point for discussions, or as a source of ideas for simple investigations. For example: what happens when an electric fan is blowing on a Goofy Bird? What happens to a Goofy Bird in a very cool place? And, what happens when a Goofy Bird performs in an increasingly more humid atmosphere, such as under an inverted aquarium (with some wetted blotting paper) in a sunny window?

Measuring the Muller–Lyer optical illusion

Alan Ward, Saint Mary's College, Cheltenham

Draw two identical straight lines with arrowheads on both their ends. But on one line draw the arrowheads inside out (Figure 1). Now one of the equal lines looks longer than the other. This is of course the famous Muller–Lyer optical illusion, to be found in most puzzle books for children. Can the 'degree of error' be measured?

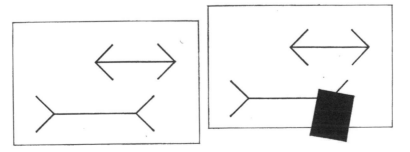

| Figure 1 | Figure 2 |

We began with an enormous diagram in which the straight lines to be compared were each 45 centimetres long. We also used a 15 cm × 20 cm black card and a ruler. Then we asked each member of the class, in turn, to cover the 'longest-looking' line with the card—so that it appeared equal in length with the other one (Figure 2).

Measurements of the uncovered portion of the line were recorded on a bar graph (Figure 3). In a group of thirty Saint Mary's students, no less than seven individuals left only 40 centimetres of the line uncovered. Four students claimed that both lines looked equal. The overall error was about 9 per cent.

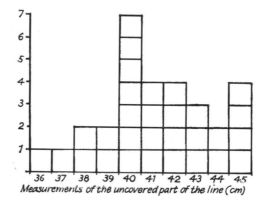

Figure 3

The activity can be presented conveniently as a 'side-show' during a routine double teaching period. More information on the Muller–Lyer illusion is to be found in *Eye and Brain*, by R. L. Gregory (World University Library, 1966), and ideas for other experiments will be suggested by S. Tolansky's *Optical Illusions* (Pergamon, 1964).

A political barometer

M. Thomson, Suhum Secondary Technical School, Eastern Region, Ghana

Teachers may be interested to know of a rather light-hearted chemical 'political barometer' which may be suitable for school open days, especially when the local MP is in attendance.

The system used is an adaptation of the famous 'blue bottle' described by Campbell [1]. A solution is made containing 20 g sodium hydroxide and 20 g glucose per litre. To this are added a few drops of 1 per cent methylene blue solution and a few drops of phenolphthalein indicator; the solution is then placed in a stoppered flask. The methylene blue is rapidly reduced, leaving the pink colour due to the phenolphthalein in alkaline solution. When shaken however, the blue colour of the methylene blue is restored and this masks the pink colour. The blue colour soon fades to leave the pink which in turn fades to leave an orange solution.

The story that goes with the political barometer follows these lines: 'The political barometer starts off being Socialist (pink), but when agitated (by shaking) it rapidly turns Tory (blue). When left alone it again becomes Socialist, but if allowed to stagnate it will become Liberal (orange).'

REFERENCE

1. Campbell, J. A., 'Kinetics—early and often', *J. Chem. Educ.*, 1963, **40**, 578.

Why balloons go down

C. Smith, Manchester College of Education

The following dramatic demonstration of diffusion is both convincing and interesting to the middle school pupil. Furthermore, the impact of bursting balloons leaves an unforgettable impression.

Some strong ammonia solution is placed in a deep container and an inflated balloon

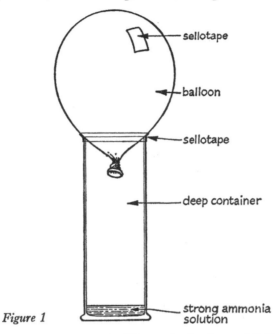

Figure 1

Sellotaped to the top as shown in Figure 1. A piece of Sellotape is fixed on to the membrane and the arrangement left in a fume cupboard for about one hour. The balloon is then removed from the container and, in another part of the room, carefully pierced with a needle through the Sellotape. With any luck, the balloon skin will not rupture with an inevitable bang. A piece of red litmus placed near the hole immediately turns blue. If a control is set up, without the ammonia solution, and tried, nothing happens, showing that the gas has passed through the membrane.

Needless to say, some balloons burst, thus adding to the interest, so it is necessary to prepare several at the beginning.

It should now be obvious why inflated balloons go down in a day or so.

An aid for reading liquid levels[1]

B. Crawshaw, Carnoustie High School, Angus

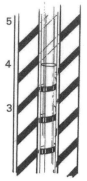

Clear liquids are hard to see in a manometer or sight tube. A background of diagonal stripes is a great help. There's some good physics, too.

An aid for drawing tangents to curves

R. M. Garrett, Beech Hill High School, Luton

I was interested in the note by Crawshaw [1] as I have been using a similar trick for finding the tangent to curves, particularly graphs. A small length of capillary tubing laid on the curve produces disjointed lines in the tubing, Figure 1(*a*), and only when the tube is exactly at right angles to the curve are all the lines of the graph and those in the tubing smooth and continuous: Figure 1(*b*). By marking points at the extremities

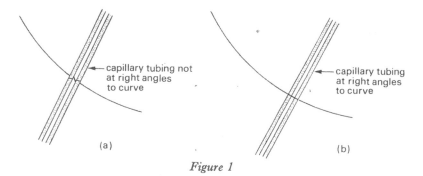

Figure 1

of the tubing, a right angle to the curve at any point may be drawn and the construction of the tangent to this is then a simple matter. It seems to work just as well with the transparent plastic barrels of some throw-away ball-point pens.

REFERENCE

1. Crawshaw, B., 'An aid for reading liquid levels', *S.S.R.*, 1976, 202, **58**, 114.

Capillary absorption

M. L. Allan, Huddersfield Polytechnic

The following technique can be used to demonstrate capillary absorption by porous materials. The demonstration works well on the overhead projector, but as the materials required are inexpensive it should be possible to provide class sets of apparatus. Comparisons can then be made of capillary absorption rates of, say, different types of brick. The difference between absorption across and along the grain of a cube of wood can also be studied.

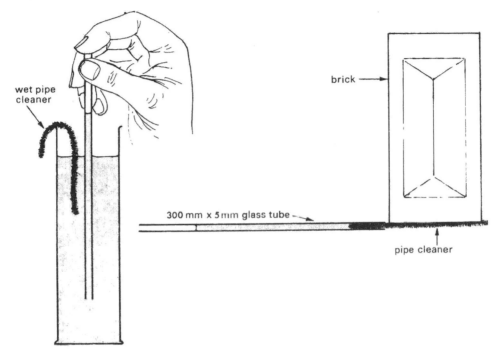

PROCEDURE

A glass tube (or transparent plastic drinking straw) is withdrawn from a cylinder containing coloured water while the top is sealed with the forefinger. The open end is inclined slightly upward while a wetted pipe cleaner is pushed 2 cm down into it. (This should prevent any air bubbles being trapped by the cleaner.) The tube and cleaner are placed on an impervious surface and the position of the meniscus marked. The test sample is placed on the pipe cleaner, close to the tube. The rate at which the meniscus moves along the tube is noted. With fairly porous materials rates are satisfyingly rapid.

A demonstration using the OHP – and a possible danger

DL Thompson

The dangers associated with focussing the sun's rays with a telescope or converging lens are well known. On a clear day it is possible to light a match or a piece of paper with a suitable lens.

A similar effect and a dramatic demonstra-

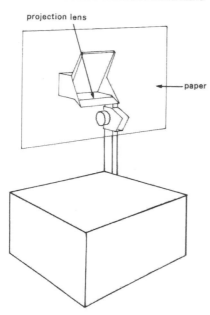

projection lens

paper

tion can be achieved with an OHP – and a nasty skin burn for the unwary. Hold a sheet of paper very close to the projector lens (see figure) and adjust the OHP to obtain a concentrated spot of light. If a match head is now held in the concentrated beam it will ignite in approximately half a minute.

DL Thompson, Faculty of Education, Huddersfield Polytechnic, Holly Bank Road, Huddersfield HD3 3BP

A simple cloud chamber

Colin Siddons

Take a round-bottomed flask, preferably one as big as 0.5 litre. Swill a little water around inside it, empty the water out, leaving the inside wet. Blow into the flask (Figure 1). Nothing happens. Suck out. A faint cloud might just be noticeable. Next drop a burning match into the flask. Suck out again. A much thicker cloud will now appear (Figure 2). It will show up even more prominently if a beam of light is shone across the flask.

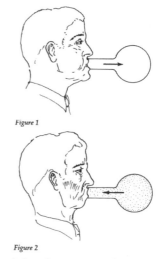

Figure 1

Figure 2

The cloud chamber, invented by CTR Wilson in Cambridge enabled the tracks of alpha particles through air to be photographed, their lengths to be measured and their collisions to be followed. The existence of positive electrons as well as the usual negative ones was shown for all to see.

Another name for the Cloud Chamber is the Expansion Chamber. When some of the air in the flask is sucked out the remaining air expands and therefore cools. If the expansion is sufficiently large the moist air will become supersaturated with water vapour. The flame of the match ionizes the air, some molecules become charged. These charges act as starting points, nuclei, for the formation of tiny droplets of water. Thus the cloud is formed.

This demonstration is not original, I saw Professor Lawrence Bragg show it in Manchester in 1938. That a burning match ionizes air is easily shown. Hold one near a charged electroscope: a total loss of charge is soon registered.

JC Siddons, 38 Oakleigh Avenue, Clayton, Bradford

A simple universal coupling

P. Watkins, Charles Keene College, Leicester

It is often necessary to couple the shaft of a small electric motor to that of a gearbox which is not in coaxial alignment with it. Thickwall rubber tubing can be used, but experience shows that this material soon shows signs of wear, and, in any case, it is difficult to neatly clamp the tube to the shafts.

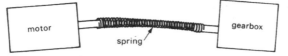

An alternative is to use a spiral spring which is a close fit on each shaft, and, importantly, tightens when the shafts rotate. The figure shows an arrangement which has been used to couple small electric motors to Meccano drives in a large model designed to test the electronic circuit of a prototype Metal Cutting Saw. (Meccano universal couplings have far too much backlash, and are incompatible to the diameter of the motor shaft.)

ADVANTAGES
1. The arrangement accommodates shaft misalignment.
2. Silent and smooth running, even at high speeds.
3. Can be fitted between shafts of differing diameters.
4. The arrangement is small and compact.
5. The spring transmits a smooth torque build-up to the gearbox shaft, therefore reducing any tendency to jerky/stick-slip motion.

Beginning to imagine the size of outer space -

instructions that you can photocopy for immediate use with your students

Alan Ward

Many years ago a magazine offered as a competition prize 'a penny for every mile from the earth to the moon'. Those were the days of 240 old pennies in the pound. The prize was a thousand pounds – so (roughly): How far away is the moon? Distances between the earth, moon, planets and stars are very difficult to imagine.

In this chart, I have given you some simplified proportions, which will enable you to make a scale model, illustrating the immense space between the earth, moon and sun. The diameter of the earth is taken as the basic unit for the model (1). *Let this unit be 1 cm.*

Diameter of the earth	7 900 miles	1
Distance to the moon	240 000 miles	1 x 30
Diameter of the moon	2 200 miles	0.35
Distance to the sun	93 000 000 miles	1 x 12 000
Diameter of the sun	865 000 miles	1 x 109

Prepare two little plasticine or Blutack balls, measuring 1 cm and 0.35 cm in diameter. Pin these balls to the tops of two sticks. They will represent the earth and moon. Make a 109 cm diameter cardboard disc, to represent the sun - and pin this to a third stick.

Erect these sticks in a field, with the 'moon' 30 cm (a scale distance representing 240 000 miles) from the 'earth', and the 'sun' 12 000 cm or 120 metres (a scale distance of 93 000 000 miles) from the 'earth'. Put the sticks in a straight line - *but you are going to need a big field!*.

Seen by putting an eye next to the model earth, your moon and sun look to be about equal sizes, as the real moon and sun appear to be...

Take time to let the full meaning of this activity make an impression on you. Doesn't this experience make your planet earth in space seem fantastically small?

On the same scale, Pluto, the sun's most distant planet would be another (0.2 cm diameter) ball, on a stick placed 4.75 kilometres away!

REFERENCE
Ward, A, *Sky and Weather*, *Science Discovery* series (Franklin Watts, 1992).

Alan Ward, 4 Branch Hill Rise, Charlton Kings, Cheltenham, Glos GL53 9HW

Air is a real substance . . .

Alan Ward, College of Saint Paul and Saint Mary, Cheltenham

It must be appreciated that air is a 'real substance'. But air, being invisible, is known by the indirect observation of its effects. Therefore we must be able to appreciate that air has mass; meaning that it occupies space, and is attracted by gravity—an effect we generally describe as weight. We should be able to show juniors that air takes up space and is capable of being weighed, but it is doubtful that young children will be able to grasp the significance of mass. These ideas can be explored with a simple piece of apparatus.

Impale a small clipboard clip A on a 5 cm nail. Fix the nail, pointing sideways, into a cork. Insert the cork in a heavy bottle. Grip a 1 m long cm² section balsawood strip in the clip, near one end of the wood. Dimensions given are not critical. Then tape a bent wire 'hook' B at the longer end of the strip, and attach some clothes pegs, or modelling clay C, on the shorter end, so that the strip roughly balances. Use a cardboard balloon pump to inflate a *very large* round balloon, and seal its neck with a 'twist' of light wire.

Stand the bottle-and-balsa 'balance' on a stool in an area where there are no draughts. Suspend the big balloon by hitching the wire upon the hook. Adjust the balance of the apparatus, until the balsa 'beam' is horizontal. Children might like to suggest alternative

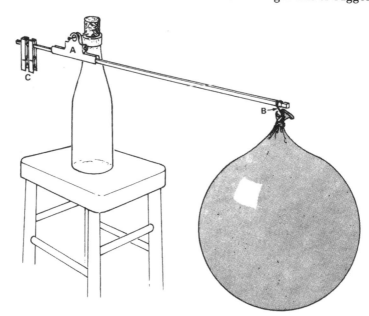

ways to get it to balance. Inevitably the balloon will have acquired a charge of static electricity. This nuisance can be turned to advantage, through discussion of the difficulty with the children. Next, the balloon must be removed, without disturbing the beam.

While the balloon is being deflated, a child can support the beam, using a finger. Afterwards, the balloon envelope is hung back on the hook by the same wire. Then, when the child's finger is removed, the beam will be seen to be slanting, with its longer balloon carrying end somewhat raised. Clearly the balloon, having lost the air it enclosed, is now less heavy than before. It has lost weight, proving that it must have 'had weight' when the experiment began. Of course the test is repeated several times.

The value of the apparatus lies mostly in the questions it provokes. Why did we use a balsawood beam that was much longer on one side? Perhaps our young scientists will object that the experiment proves nothing, because the weight of the air inside the balloon is equalled by the buoyant upthrust of the air around it. (It might be hoped that very young children would not notice this anomaly . . .) Of course, the mass of air we have investigated was the difference between *compressed* air inside the balloon and the less dense air outside.

The bicycle—a useful resource

J. W. Hawley, Inkersall Junior School, Derbyshire

Primary teachers must rely on everyday objects for resources to base their science lessons on since funds rarely allow primary schools to purchase expensive equipment. This is true particularly for apparatus concerned with studies about man-made artefacts and mechanical actions. The bicycle, however, is a valuable resource and one which is readily available. Most classes have a member who is prepared to bring his/her bicycle into the classroom. A model with derailleur gears is best suited for this purpose.

The following notes are some ideas for science lessons arising from consideration of a bicycle:

Structures
The frame is basically made up of two triangles which are strong shapes. Children can test this idea using meccano.

Forces
Consider pulling, pushing and twisting forces, e.g. pulling brake lever, pushing down on pedal, twisting the handle-bars.

Friction
Areas of low and high friction can be identified, e.g. wheels turn easily (low friction), brake blocks 'rub' on wheel rim (high friction).

Materials
Bicycles are made up of metal, rubber and possibly plastic components. Why do the properties of each material make it suitable for the components function e.g. strong metal frame, elastic rubber tyres, light plastic mud guards?

Movement
Consider the circular movement of the pedals, chain, rear sprocket and wheels.

Speed
The distance/time relationship can be studied following 'time trials' on the playground.

Gears
Count the teeth on each cog. What effect does this have on the distance travelled per revolution of the pedals?

Levers
Brake levers, gear levers, the steering mechanism and the pedals are all examples of levers.

Air
Consideration of the pneumatic tyre leads on to pumps, valves and air pressure.

Electricity
The cycle lamp provides us with an example of an electric circuit. The power source may be a battery or a dynamo.

Light
Whereas the lamp is a source of light the reflector uses light from another source.

Sound
The bicycle may have a hammer operated bell or an electric buzzer.

No doubt the reader can think of further possibilities for studies arising from consideration of the bicycle.

Electric writing

John Mattocks, Bristol Polytechnic

The following experiment is fun, easy to do and instructive.

Soak a piece of filter paper, or white blotting paper, in potassium iodide solution. Remove the paper and shake off the excess solution leaving the paper damp, but not soggy. Spread it out on a sheet of tin plate such as a large tin lid. Connect the lid to one terminal of the *low tension* supply of a lab pack and a *defunct* ballpoint pen to the other. If the ink tube is removed from the pen, a thin copper wire can be threaded down the barrel and can be nipped between the case and the brass end. Now write on the paper with the lab pack switched on to 6 V d.c. Try again with the polarity reversed and again on 6 V a.c. (avoid touching the tin with the pen point to prevent blowing a fuse).

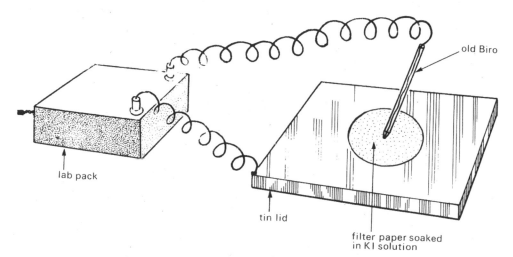

old Biro

lab pack

tin lid

filter paper soaked in KI solution

Figure 1

There is a number of questions that can be asked such as:

What makes the brown line appear?
Why does a mark appear with one polarity and not the other?
   (The answer to this is not as simple as it at first appears to be!)
Why is the a.c. line dotted?
What does the interval between the dots represent in time?
Are the dots and intervals of equal length? Why?
How could you use the principle to measure the mains frequency and, given the frequency,
   to measure the revolution speed of a record player turntable?
Why do the brown marks fade after a while?

Pivots for science apparatus

John Mattocks, Bristol Polytechnic

One of the problems met in making science apparatus for use in schools is the provision of a cheap, robust, low-friction pivot. In relevant literature the female part of a snap fastener is often suggested. I have never found this to be entirely satisfactory. An old ball-point pen and a combustion tube are much more successful and cost little.

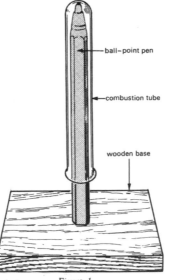

ball–point pen

combustion tube

wooden base

Figure 1

A combustion tube of internal diameter a little wider than the pen is needed. The pen is sawn off about 2 cm longer than the tube and is glued into a hole in the centre of a wooden base. As an alternative to the tube, a length of glass tubing which has had one end sealed by rotating it in a bunsen flame is equally effective.

Suggestions for the kind of apparatus for which the pivot is useful include a compass Figure 2(a), a fan for convection currents Figure 2(b) and a device to show electrostatic attraction and repulsion Figure 2(c).

For the compass, two discs of cardboard are cut out and old razor blades which have been magnetized are glued between them. A hole is made in the centre of the discs with a cork-borer of diameter appropriate to achieve a push fit on to the combustion tube. Obviously it is necessary to align the razor blades so that they are parallel and have their polarities in one direction. The magnetic axis can be marked on the upper disc with a felt pen.

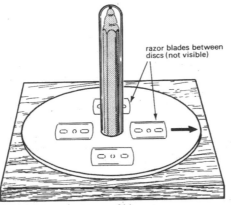

razor blades between discs (not visible)

2(a)

In the electrostatic apparatus a plastic knife is fixed to the tip of the combustion tube with a dab of glue, or with a loop of sellotape. Plastic spoons will work, but are not quite as long as the knives.

For the convection fan, instead of using a wooden base, the pen can be held in a retort stand and so a low bunsen flame, or a candle can be used as a source of gentle heat below. The friction of the pivot is so low that rotation can be achieved even with the heat from a 60 W light bulb.

No doubt readers can device other uses for this practical pivot.

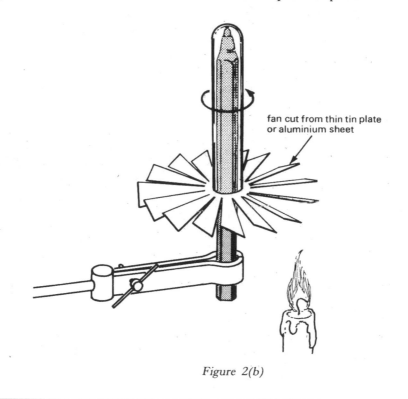

fan cut from thin tin plate or aluminium sheet

Figure 2(b)

A 'magic' bubble motor

Alan Ward, College of Saint Paul and Saint Mary, Cheltenham

The picture shows an effective bubble motor that I have just made up. It is mounted on a 'pillar' made from a Domestos bottle with its top cut off, on top of which is a saucer, to support an inverted milk bottle. (The 'dip' in the milk bottle's bottom is convenient.) Supporting arms for the motor are formed from about 60 cm of stiff iron wire. The 'motor' itself is a yogurt pot with a hole (to receive a pea-shooter) bored approximately 3 cm from the opening. Just before use, it is hung, upside down, by attaching it to a short roll of BluTak fixed to an end of the wire. I have used a little padlock as a counterweight, hooked to the other end of the wire.

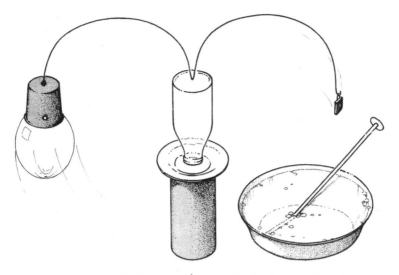

To prepare the motor, attach the pea-shooter, dip the inverted pot in a strong solution of liquid detergent, and blow a large dangling bubble. Carefully pull out the pea-shooter, keep a thumb over the hole, and hang the pot-and-bubble on the wire—making sure that the hole (which I made with a chemist's cork-borer) is pointing 'along a tangent' to the potential circle of rotation of the device. Gently let go, taking care to dampen any vibration of the wire—and wait a moment. Ever so slowly at first, the motor starts to drive itself—by jet-propulsion—and it gathers speed, as the bubble gets smaller (and pressure of the air inside increases) [1]. It should completely circumnavigate the milk bottle.

REFERENCE

1. Ward, A., 'An unusual parallelogram of forces', *S.S.R.*, 1976, 203, **58**, 335.

Seeing 'ghosts'

Alan Ward, College of Saint Paul and Saint Mary, Cheltenham

Isa Bowman was the child actress who starred in a production based on the story of *Alice in Wonderland,* performed at the Royal Globe Theatre in 1888. She remembered the author, Lewis Carroll, showing her a book of ghosts when she visited him in Oxford. The ghosts could be made to appear on ceilings and walls. Her favourite ghost was a green picture that appeared as a pink coloured spectre in Lewis Carroll's rooms.

Photosensitive rod cells in the retina of an eye contain a chemical called rhodopsin, or 'visual purple'. Exposed to light, rhodopsin is bleached, first yellow, then a colourless mixture of opsin and Vitamin A. This chemical decomposition, brought about with energy from photon particles in the light ray, triggers nerve pulses that alert the brain, where the light stimulus is perceived—and we 'see'. The rod cells affected remain insensitive, or 'fatigued', until the visual purple is restored by recombination of Vitamin A with opsin. For this to happen, the cells need a period of relief, or rest, from exposure to bright light. The cells become most sensitive after some time in the dark.

Rod cells respond to light in a general way. Cone cells, concentrated in the part of the retina called the fovea, where we usually focus our eyes, react in a similar way to rod cells—but it is thought that there are three kinds, each sensitive to what physicists call the primary lights: red, green and blue. According to a splendidly successful theory, various intensities of red, green and blue lights can produce any colour that we see, the most dramatic example being colour television. Observe how 'natural' colours are produced on the screen by three intensities of a thousand sets of primary coloured dashes.

Interesting evidence, supporting the theory of colour-vision, is obtained by staring at patches of primary-coloured papers stuck on white cardboard. After twenty seconds of staring at a red patch, look at the bare whiteness alongside. There, *apparently,* a pale greenish (turquoise) so-called after-image is perceived. Why does this 'ghost' appear? Cone cells that were activated by red light reflected from the paper patch have become temporarily fatigued, losing their sensitivity. But cone cells that are sensitive to green and blue have been resting—so they are subsequently stimulated by green and blue light reflected by the white surface. (For the purpose of this explanation daylight is considered to be a mixture of red, green and blue lights.) The blue and green lights produce an impression of blue-greeness in the mind, hence the ghost.

After-images to green and blue patches are red-blue (magenta) and red-green (yellow). A yellow paper patch reflects both red and green lights, fatiguing red and green cone cells, to produce an after-image in blue. Isa Bowman's 'pink' ghost was really pale magenta.

During the children's happenings associated with the 1982 Cheltenham Literary Festival, I gave a presentation on 'The Puzzles of Lewis Carroll'. I showed 'Isa's ghosts' to two hundred eleven year olds. I used three large white cards, each decorated with one of three cut-paper images: a smiling black imp, the green face of a rabbit, and a grinning yellow skull. Staring at the black figure fatigued all three kinds of cone cells, to produce a greyish after-image. This was apparently visible on a plain white card with which I covered the picture—after the children had stared at it for twenty seconds. About half the children saw the grey ghost. There were gasps of wonder, when more children perceived the pink ghost of the green rabbit, and there was a stir of excitement when everybody saw the blue ghost of the yellow skull. (The reader could adapt the idea by making coloured transparencies for a slide-projector.)

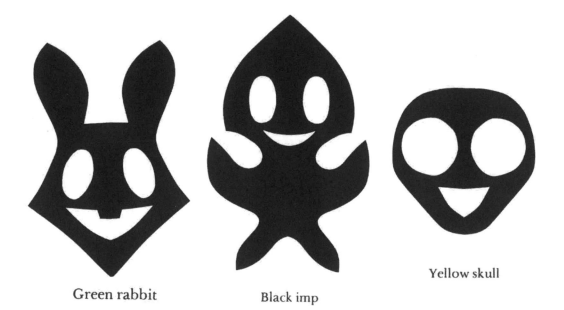

Green rabbit   Black imp   Yellow skull

Shifting equilibrium—the incredible Japanese 'tipper'

Alan Ward, Formerly of the College of Saint Paul and Saint Mary, Cheltenham

Here is a rare piece of Japanese origami (paper-folding) that provides a fascinating lesson in mechanics. Always make up a fresh model for a demonstration. Start by folding in half a thin paper square (22 × 22 cm) along one of its diagonals. Bend in two of its adjacent sides, to meet along the diagonal—and make the kite shape (Figure 1).

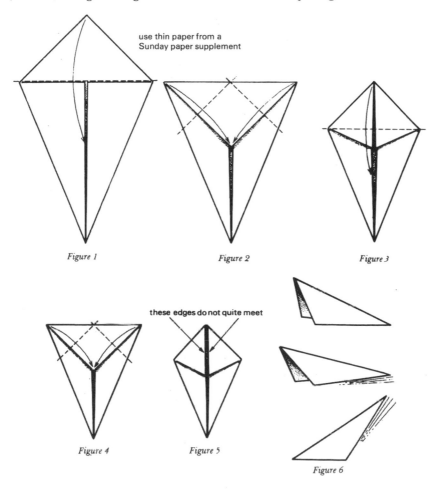

use thin paper from a
Sunday paper supplement

Figure 1   *Figure 2*   *Figure 3*

these edges do not quite meet

Figure 4   *Figure 5*

Figure 6

Thereafter, carefully follow the order of the drawings, until you get the smallest kite shape (Figure 5). The bundle of paper will feel fat at this stage. Notice that the parts of the top edge from Figure 4 do not quite meet in the middle on Figure 5. *This will help to prevent the paper splitting* when, finally, you fold the small kite shape across its mid-line, to make the sphinx like object that you see—resting on a smooth, flat surface—at the top of Figure 6.

Now watch . . . The tail end of the object begins to quiver and rise. It moves more rapidly (as its front end splays outwards) and—probably sooner, rather than later—it suddenly tips up, to resemble a witch's hat.

Some energy is stored in the springiness of the two flaps inside the device. When these flaps begin to spread, they push the base of the object outwards.

The effect is to shift the centre of gravity forwards, until it is no longer covered by the base. Then gravity pulls the thing upright. Play with the idea. You will learn not to squash it too flat before use. Adjustments to the springy flaps might be needed. Experiment to gain expertise. Engineers will be enthralled. Invent a proper use for the device. Tippers that take a little time to work are the most weirdly entertaining.

Use a syringe for an instant vacuum

Michael Kahn, University of Botswana

A neat way of showing that sound does not pass (easily) through a vacuum is provided by using a syringe. The technique described here works for both glass and plastic syringes. The best vacuum results with a large glass syringe, but a 25 cm^3 plastic one is capable of giving a vacuum down to 30 kPa if the plunger is rapidly withdrawn while the nozzle is closed with a finger.

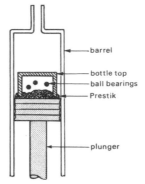

The only item now needed is some kind of rattle which can be enclosed in the vacuum. This can be improvised by attaching a small plastic bottle top to the rubber piston by means of 'Prestik'. Three small ball bearings are placed inside this top to act as a rattle. Naturally the top must be small enough to enter the barrel, and this clearance and the damping of the Prestik and piston ensures little sound is transmitted to the outside directly through the walls of the syringe by transmission (see figure).

Evacuate the syringe. Shake well. Little sound is heard. Remove finger. Whoosh. Shake. Lots of sound is heard.

A fresh look at 'dancing mothballs'

Alan Ward, Formerly of the College of Saint Paul and Saint Mary, Cheltenham

Old books about scientific recreations mention the 'dancing mothballs'. Half a teaspoonful of sodium bicarbonate is added, a little at a time (to prevent an uncontrollable effervescence) to water in a drinking glass that already contains a teaspoonful of dissolved white vinegar. The reacting chemicals set free copious quantities of carbon dioxide gas. Then, after the fizzing subsides, four or five mothballs are added.

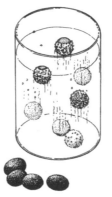

Being slightly denser than water, they sink. But their surfaces provide nuclei for attracting gas. Soon they are covered with silvery bubbles that lift them to the top of the water, like little balloons. There the mothballs rock and roll. This is because, as bubbles swell or break, the forces of flotation vary unevenly. Eventually the mothballs lose critical amounts of buoyant gas, so they sink—but they regain bubbles, to rise again and perpetuate a magical up and down 'dance'.

Years ago I could never get the receipe to work reliably, so I put a quantity of dilute hydrochloric acid into tap water with the sodium bicarbonate. I used to put a large beaker of dancing mothballs on display at laboratory Open Day exhibitions, adding an extra attraction by using mothballs coloured with wax crayons.

However I was uneasy about telling children my impatient trick with the hydrochloric acid. Last summer I devised a delightfully safe and cheap way to present the mystery of the dancing mothballs. Use a glass of fresh clear lemonade. After half an hour of being buoyed up by carbon dioxide gas released from fizzy lemonade, mothballs start to dance . . . The trouble nowadays is that mothballs are virtually unobtainable.

Two years ago a friend told me about an ancient shop in the country, where an old-timer shopkeeper had an enormous supply of mothballs. So I bought 4 kilos. I keep them double-wrapped in two plastic bags. But enough of this anecdote. I have found that grapes work well. Also try sultanas. It might be thought to be a pleasant scientific task for children, to find other objects of suitable densities, to stand in for the traditional mothballs.

A statistical analysis of paper helicopters

Kevin S Barber and CJ Sweeney

In response to curiosity sparked by an article [1], a statistical analysis was performed on the performance of paper helicopters. Extensive data was collected on the flight time of various helicopters to see what effect a variation in size and weight had on the flight time of the helicopters. A total of 26 helicopters were made, following the design introduced by Alan Ward in his article [1]. The helicopters were tested with four variations in size, and respectively, in weight. The dimensions of the helicopters ranged from 2 × 9 cm to 12 × 40 cm, with masses from 0.20 g to 3.61 g. Four helicopters for each of four sizes were made and dropped eighty times from a height of 2.44 m. The flight time of each helicopter was timed with a digital stopwatch.

Heli-copter	Size(cm)	Total area(cm²)	Wing area(cm²)
A	2 × 9	18	9
B	3 × 10	30	15
C	6 × 20	120	60
D	12 × 40	480	240

Heli-copter	Mass(g)	Mean flight time (s)	Standard deviation(s)
A	0.20	2.96	0.13
B	0.29	2.62	0.20
C	0.99	2.48	0.21
D	3.61	1.46	0.13

(The weight, mean flight time, and standard deviation of helicopters A, B, C and D is an average of the four helicopters.)

The results show how the flight time of each helicopter varies according to the size of the helicopter, the mass of the helicopter, and the wing area. The statistics show that the larger the total area of the piece of paper, the shorter the flight time, while a smaller total area results in a longer flight time. The same relationship can be seen between the flight time and the mass of the helicopter.

REFERENCE

1 Ward, A, 'Exploring air resistance with paper helicopters', SSR, 1975, **57** (198), 140.

KS Barber and Fr CJ Sweeney, St John's High School, Toledo, Ohio, USA

Lilly's Formula for radioactive decay

Richard Weston

Some years ago a sixth former, Simon J Lilly, pointed out to me that the conventional methods for solving problems on radioactive decay were unnecessarily cumbersome. It was in the days when the electronic calculator had just come in with its 'x to the power y' function.

Faced with a problem such as:

What fraction of a sample of radioactive carbon-14 will remain after ten thousand years if its half-life is 5570 years?

most teachers and some pupils would respond

'*Well . . . about a quarter . . .*'   and then try to remember the formula: $\log(N_o/N) = kt/2.303$ (where N_o is the original number of atoms, N is the number of atoms at time t, and k is the first order rate constant which can be shown to be equal to $0.69/(\text{half life})$) [1].

All this was rather taxing on the memory and probably only those pupils with a good grasp of the calculus were convinced when they were shown the derivation of these formulae.

Consequently, before Lilly came along, most of my A-level chemistry students used to find such calculations somewhat mysterious. Also in examinations such problems proved to be time consuming.

I now will describe Lilly's reasoning which I have used in my teaching for the last ten years and which pupils seem to find natural and easy to grasp, even early on in an A-level chemistry course.

Pupils can be introduced to the concept of the constant half-life by measurement of the half life of protoactinium-234 [2] and by demonstrating a computer simulation of radioactive decay[3].

The exponential decay curve is then drawn and explained.

It is then easy to see that :
After one half-life, fraction remaining = $(1/2)$
After two half-lives, fraction remaining
$$= (1/2) \times (1/2) = (1/2)^2$$
After three half-lives, fraction remaining
$$= (1/2) \times (1/2) \times (1/2) = (1/2)^3$$
And after n half-lives, fraction remaining
$$= (1/2)^n$$

This is LILLY'S FORMULA:
Fraction remaining = $(1/2)^n$, where n is the number of half-lives which have elapsed.

Now $n = t/t_{1/2}$, where t is the time elapsed and $t_{1/2}$ is the half-life
Lilly's insight was to realize that n need not be a whole number and that the formula must work for all values of n including fractions.

Thus we have a very simple, easily memorized, understandable formula for radioactive decay which can readily be applied to solve the standard types of problem set in A-level chemistry examinations.

EXAMPLES

1 *Using the problem quoted above:*
number of half-lives, n
$$= t/t_{1/2} = 10000/5570 = 1.795$$
Therefore, fraction remaining
$$= (1/2)^n = (1/2)^{1.795} = (0.5)^{1.795} = 0.288$$
(The last step of the calculation is easily carried out on the calculator by using the 'x to the power y' button with x = 0.5 and y = 1.795).

Thus after 10000 years the fraction of carbon-12 which remains is 0.288.

2 *How long will it take for 99% of a sample of Co-60 to decay if its half-life is 5.23 years?*
Fraction remaining = $(1/2)^n = 0.01$
taking logarithms: $n\log(1/2) = \log(0.01)$
$n = \log(0.01)/\log(0.5) = 6.64$.
The time required for 6.64 half-lives to elapse will be $6.64 \times 5.23 = 34.7$ years.
It will take 34.7 years for 99% of a sample of Co-60 to decay.

3 *What is the half-life of a radioactive isotope which decays to 0.397 of its original activity in 60 seconds?*

Fraction remaining = $(1/2)^n = 0.397$
taking logarithms:
$$n = \log(0.397)/\log(0.5) = 1.33$$
If 1.33 half-lives correspond to 60 seconds then,
one half-life = $60/1.33 = 45.0$ seconds.
The isotope's half-life is 45.0 seconds.

REFERENCES

1 Nuffield Advanced Science, *Chemistry*, Book II, (Penguin Books, 1970), p 62.
2 Nuffield Chemistry, *The Sample Scheme*, (Longman, 1966), p 445.
3 Sparkes, RA, *The BBC Microcomputer in Science Teaching*, (Hutchinson, 1984), p 83.

RG Weston, King's School, Bruton, Somerset BA10 0ED

Talk 'science' as you play with a candle motor...

Alan Ward

People will be entertained by this modified form of the ancient candle motor. It gives a focus for discussion about mass, weight, force, moments, levers, and the physics of see-saws – all in an atmosphere of fun. Children encouraged to build their own machines must be warned about the dangers of playing with matches and fire. Demonstration by a parent or teacher might be the preferred way to present this item, which is fine if there is also thought-provoking dialogue with the children.

Insert a steel knitting needle at right angles across the middle of a plastic straw. Fit a birthday-cake candle into each end of the straw, but if the candle ends are too fat, soften by dipping into hot (not boiling) water, before rolling them between your fingers. Rest the points of the needle on the bottom of a pair of upturned cups. Everything should be assembled on a

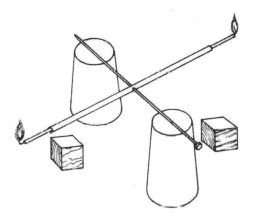

metal tray. Light the candles.

More mass of wax always drips off the lowermost candle, causing irregular displacement of the system's centre of gravity. Loss of mass is clearly connected with loss in the force of weight. Lever law turning moments on either side of the needle fulcrum alternately exceed each other in their effects. (Put wooden blocks underneath the straw's ends, to stop the candles from touching the tray.) Blow out the candles before they start to melt the plastic – and generate toxic gas... The charmingly erratic see-sawings of this candle motor look especialy attractive in an otherwise darkened room. **Play safely!**

Alan Ward, formerly of the College of Saint Paul and Saint Mary, Cheltenham.

A kitchen-sink hovercraft

Colin Siddons

After washing with soapy water a wine glass is swilled out with very hot water. It is then placed upside down on a flat wet draining board which has a slight slope.
What happens? Why?

The glass slides down the slope. Small bubbles appear *outside* the rim of the glass at first but after a time they turn round and go *inside* the rim where they become more plentiful (see Figure).

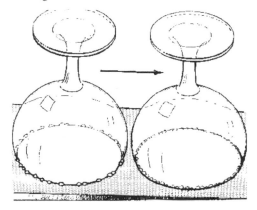

The hot water left inside the glass evaporates quickly. As the water is very hot the pressure produced by its vapour is high. The total pressure inside the wine glass becomes greater than the pressure outside. This lifts the glass slightly allowing some air to escape. It becomes a hovercraft. It will slide down even a very slight slope.

But the air inside the glass is cooling rapidly and so its pressure decreases. The effect of this cooling opposes the effect of the evaporation. Eventually cooling predominates over evaporation. This is the moment when the bubbles turn round and go back into the glass.

JC Siddons is now retired, having previously taught at Thornton School, Bradford

A problem-solving ploy with a party popper

Alan Ward

Wanting a cheerful climax for a primary schools in-service training day on 'Problem solving across the curriculum', I set myself the task of firing a Party Popper by means of a fuse and detonator consisting of dominoes. Soon I met two difficulties:

1 How would I be able to apply a sufficient force on the popper's firing string, to make it explode?
2 How would I keep the popper fixed in one place, while the force is being applied to it?

To create a rousing effect, I wanted to fire the popper vertically, so that its brilliantly coloured paper streamers would appear in the air above my table.

I solved the first sub-problem by using a thin strip of wood, a long rubber band and a large size paper clip. These were used to grip the popper, to point upwards, at the edge of a tall table. Eventually I solved the second sub-problem with a heavy steel ball attached to the popper's firing string by a length of strong thread. It took me some time to work out how long the thread should be, as I needed the ball to drop some distance, before getting enough momentum to give a hard enough jerk to the firing string.

The pictures show how the popper and ball were placed side by side on the high table. A chain of standing dominoes as a 'fuse' led to a 'detonator' made by piling two extra dominoes on top of the last upright one in the chain. My intention was to get the tumbling detonator to knock the steel ball off the table – to fire the entertaining bombshell... Several trials at home worked splendidly – and my wife was so amused that she excused the mess I made!

I present the idea to teachers as a *fait accompli*, but I explain to them the various difficulties I had to made it work satisfactorily. Then I suggest that they use the idea as a problem-solving challenge to children: 'Use a box of dominoes

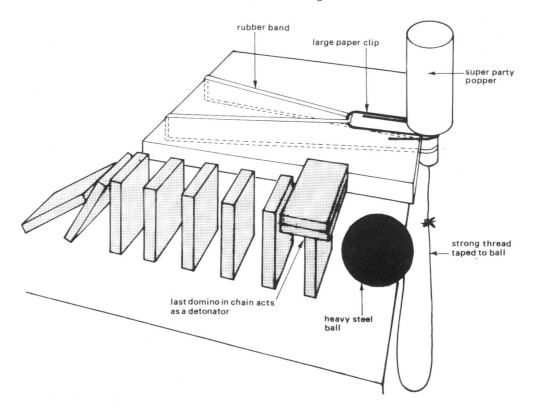

rubber band

large paper clip

super party popper

strong thread taped to ball

last domino in chain acts as a detonator

heavy steel ball

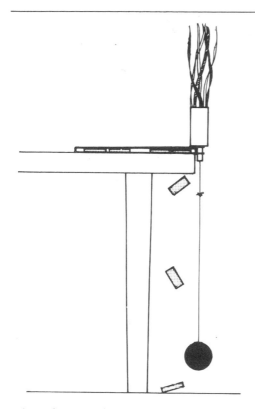

(plus whatever else you may need) to get a Party Popper to explode *safely'. Children must be warned that Party Poppers should never be pointed at people's faces!*

Later comes an ultimate challenge: 'Use as many dominoes as you like, to get a chain of dominoes on one table, to activate chains of dominoes *on another table* to fire a salvo of three Party Poppers - and raise the National Flag on a flagmast fixed on top of a flight of steps made from building blocks... (A 'Tarzan domino' will be needed to transfer energy between tables.)'

NOTE

Party poppers are made of plastic and cardboard. They are shaped like bottles, 6 centimetres tall. They contain a small explosive charge which is set off be pulling a string attached to the bottle's neck. Exported from China, they are easily obtainable from newsagents and novelty shops for about a pound for a box of twelve. They explode with a flash and a puff of smoke – and shoot a bouquet of coloured streamers into the air.

A Ward, formerly of the College of Saint Paul and Saint Mary, Cheltenham

Cherchez la Femme Francaise - a 'magic motivator' for physics...

Alan Ward

Five plastic capsules with snap-on tops - cylindrical roll-film containers retrieved from a waste-bin - gave me an idea. I also needed my smart blue clipboard, a French 1 franc coin (from a coin-collector's shop), four British 5p coins that are similar in colour, size and weight to the French money - and a little of my 'magic'. I face the children and use the back of my board as a tray, to display the capsules, their separate tops, and the coins. A child will come to me and agree that I have five empty boxes, four British coins, and a French 'lady' - as there is a young female depicted on the 1 franc. We are going to play *Cherchez la Femme Francaise.*

My assistant's task is to put a different coin into each box, to replace the lids, and to mix the boxes on my tray. Since I have just closed my eyes and turned my head aside, I genuinely do not know where the French woman is concealed. But I will discover her by magic... I test each box by shaking it (to eliminate the British-sounding rattles), by moving it on the tray, and by holding it up to my brow. There is scope for fun and nonsense, as I push four boxes to the far end of the tray, keeping one box aside. My helper will open the boxes. The four rejects all have 5p pieces. Then I win applause - when the last box contains the French coin.

How do I do it? The children offer hypotheses. We have a lively discussion. Somebody suggests we should test the coins with a magnet. Yes, the metal clip under my makeshift tray is a ready-made parking place for a magnet, before I perform the trick. It is a natural movement for my left palm to 'catch' this magnet and slide it across to a convenient spot under the tray... While I am holding the box with the French coin inside it, a magnetic pull informs my kinaesthetic sense where *La Femme* is hiding. Unlike the 5p pieces, the French coin contains ferromagnetic nickel. The children learn that not all metals are attracted to a magnet - but I urge my audience to keep the magic trick a secret.

Alan Ward, 4 Branch Hill Rise, Charlton Kings, Cheltenham, Glos

Section 7: Miscellaneous Items

Introducing the radioactive decay equation

NJ Wilson

Many A-level students do not find the equation $dN = -\lambda N\,dt$ ($dN/dt = -\lambda N$) for radioactive decay particularly easy to understand. I have used the following approach to shed some illumination on this subject.

Imagine a fly in a box with a single hole; the fly is blindfolded and flies around randomly, trying to escape. It makes numerous fruitless collisions with the walls. We can ask the question: 'what is the probability of the fly escaping?' The answer is of course 1: the fly will certainly escape (if it lives long enough). More sensible is the question: 'what is the probability of the fly escaping in a given time interval (Δt)?' We can find an 'experimental' answer as follows: look at the box while, say, 60 seconds is timed; repeat a large number of times, (N). If ΔN is the number of occasions on which it escaped, the probability of escape is $\Delta N/N$. The fly must be returned to the box when it does escape.

Alternatively, equip yourself with N identical boxes and flies and look at the collection of boxes for 60 seconds. ΔN is the number of flies escaping in that time. (Again, they must be returned to the box as soon as they escape.)

What does the probability depend on?

1 The size of the box, the size of the hole, the speed of the fly and so on, ie the physical parameters of the system, and
2 The length of the time interval.

We can next ask the question: 'what would the probability be if the time interval was increased to 120 seconds, decreased to 30 seconds and so on?' The answers: twice and half, respectively, of the probability for 60 seconds; the probability of escape changes in proportion to the time interval. It is therefore clear that the probability divided by the time interval (Δt) is constant, ie the 'probability of the fly escaping per unit time', $(\Delta N/N)/\Delta t$, is constant.

We now change the rules of the game so that the fly is not returned to its box when it escapes. As time passes, the number of boxes-with-flies decreases. We now have to take Δt small enough for N not to have altered substantially, indicated by using 'd'-notation (dn, dt). The probability of a fly escaping per unit time becomes $(dN/N)dt$, and this is constant, but the number escaping per unit time decreases, because the number, N of 'active' boxes decreases.

This is a plausible and easily understandable model for radioactive decay. The alpha (or beta) particle of course is not put back in the nucleus, so that the number of radioactive nuclei, N, decreases as time goes on (we omit at this stage the possibility of the daughter nuclei being radioactive). If we think in terms of the change in the number of active nuclei, which is numerically equal to the number of particles emitted, then dN is negative and hence

$$(dN/N)/dt = -\lambda$$

where λ is the decay constant (the probability of alpha (or beta) particle emission per unit time), or $dN = -\lambda N\,dt$.

NJ Wilson, York Sixth Form College, Sim Balk Lane, Tadcaster Road, York YO2 1UD

Fun with physics – three entertaining curiosities

Alan Ward

TO MAKE A 'K-NIBBLER'

Distort a metal coathanger as shown in Figure 1, and impale a wooden platform on the end of the hook. I used a large Meccano wheel instead of the wood. To demonstrate the k-nibbler (and I have no idea where this interesting word comes from), rest a wooden toy block on the wheel or platform, then twirl the device around on a finger or thumb. With practice, you should get the equipment to twirl, without the block falling off. The block apparently defies gravity... *Practice safely!* The activity will enhance understanding of dynamic inertia and centripetal force. I do my own k-nibbling with a large green cloth frog.

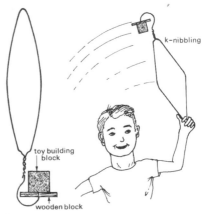

Figure 1

SUCKERS

Some polystyrene food trays can be demonstrated as 'suckers'. Experiment to find out which ones obtained locally will work best. Press together an identical pair, rim to rim, to make a pasty-like object (Figure 2). If the rims are flat and smooth, some air between the trays gets squeezed out, and elasticity of the trays ensures that there will be a partial vacuum between them - so they remain 'glued' together by atmospheric pressure. However, it is much more fun to press the trays singly against flat surfaces, such as walls and windows. Then, imperfections in the surface usually ensure that air leaks out from underneath the trays. As the plastic recovers its original form, the stiff material stutters creakily - and noisily.

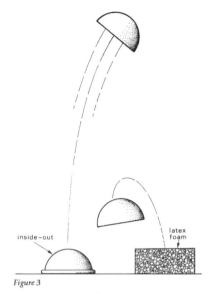

Figure 3

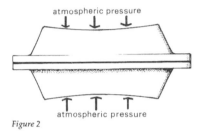

Figure 2

JUMPING POPPERS

Children will have discovered 'Jumping Poppers' (Figure 3) on sale for a few pence in newsagents and toyshops. They are small hollow hemispheres made from a type of plastic. When one of these toys is turned inside-out and put on a hard surface, it begins to assume its original shape, at first gradually, finally violently - causing itself to go 'pop' and jump over a metre in the air. I use a jumping popper to illustrate storage and release of energy. But my most dramatic demonstration is to show the principle of a shock-absorber. The energy absorber is a pill-shaped chunk of latex foam, measuring 3 cm diameter (the same as the popper) and 1.5 cm thick. The foam was used as packing material to stop a capsule of tablets from rattling. If a jumping popper is launched from on top of the latex foam, it hardly jumps 2 cm... Most impressive!

A Ward, 4 Branch Hill Rise, Charlton Kings, Cheltenham, Glos GL53 9HW

How to stand on water - scientific amusement

Alan Ward

You need seven corks, all the same size. Wine-bottle corks will do. Try to float each cork, *to stand upright* in a bowl of water. Challenge friends to try. Without 'the appliance of science' you all fail...

The trick is to arrange the corks as a flower-like raft - with all of them standing up. Keep the corks together with your fingers, while holding them underwater.

Let them go at the surface. With practice you will get the raft to float without coming apart. See Poyet's classic illustration (Figure 1) from

Figure 1

Professor Hoffmann's 1890 English translation of *Science Amusante*, by M Arthur Good. The pulls of water molecules act like elastic to hold the corks together.

All such stunts look marvellous when performed with skill.

POSTSCRIPT

When I challenged a six-year-old to get at least one of the corks to float standing up, the child tackled the problem in an unexpected way. We were floating the corks in a plastic aquarium that had curved corners. Cleverly, the child put a wetted cork into one of the corners. There the forces of surface tension and capillarity act as they do to ensure success when all seven corks are formed into a raft.

REFERENCE

Hoffmann, Professor (the pen name of Angelo Lewis), *Magic at Home - a Book of Amusing Science*, (Cassell, 1890).

Alan Ward, 4 Branch Hill Rise, Charlton Kings, Cheltenham, Glos GL53 9HW

A 'pushover' for problem solvers

Alan Ward

PROBLEM

A magician stands these sealed cigar tubes close together on a table, before covering them with a glass bell-jar. She then begins to make passes over the jar with her hands. Suddenly the black tube falls down. Can you think of an explanation?

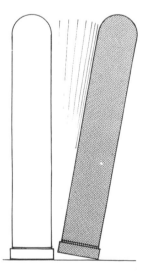

SOLUTION

Both tubes contain small cylindrical magnets (1 cm diameter × 3 cm long), but the white tube is also filled with treacle. Originally the magnet in the white tube is at the rounded end. When the trick is in progress, this magnet falls slowly through the very viscous treacle. (Viscosity is the property of a fluid whereby it tends to resist relative motion within itself.)

Soon the bottom pole of the descending magnet inside the white tube starts to repel the top (like pole) of the magnet inside the black tube. Since the total mass of the black tube is less than that of the treacle-filled white tube, it is pushed over by the force of magnetic repulsion between like poles.

USE OF THE IDEA

I call devices that work in this way 'treacle machines'. The trick with the two standing tubes is mysterious. It can be used as a stimulus to get problem solvers thinking... Just allow it to happen, before challenging observers to figure out how it works.

Other explanations are feasible and might work. You might decide not to confirm or deny hypotheses. You could then challenge your problem solvers to get their various ideas to work! My ingenious readers will surely be able to imagine and devise other applications of the general principle of 'treacle machines'...

A Ward, 4 Branch Hill Rise, Charlton Kings, Cheltenham, Glos GL53 9HW

Dotty games with dominoes - ideas for problem solving and science clubs

Alan Ward

By standing up a domino you transfer energy to it. This small amount of stored or potential energy is set free as movement, or kinetic energy, when you push the domino to make it fall over.

A falling domino can transfer its energy, to splash water, to ping musically on the rim of a brandy glass, to switch on a light, or just knock over another domino. If you stand up a closely-spaced line of dominoes in a 'chain', face-to-face, reaching across the floor, you can use them like a fuse on a stick of dynamite, to transmit a small force - that will make something interesting happen on the far side of the room.

Buy several sets of cheap wooden dominoes. Use them to test your own amazing ideas... Perhaps you can get a chain of dominoes to transfer energy up a slope or steps, to raise a paper flag on a mast at the top.

CHALLENGE

Here is a way (Figure 1) that I used a domino as a 'Tarzan domino' or pendulum, to fall sideways and transfer energy across a gap separating two domino chains standing on books.

I challenge you to adapt this method, to transmit a force across a metre-wide chasm between

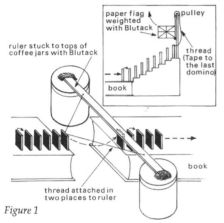

Figure 1

two tables.

The chain of falling dominoes on the other side must do something spectacular, such as exploding a mouse-trap (to throw ping poing balls in the air), or bursting a balloon.

PROPELLING A BOAT ON A MINIATURE LAKE

The boat is a polystyrene food dish (used to package meat in supermarkets) and the lake is a plastic plant-pot holder filled with water (see Figure 2).

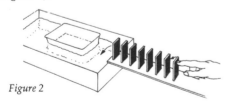

Figure 2

Each time I do this stunt I have to make sure that the boat and lake are perfectly clean, with no grease or detergent on them.

However, I do wet the top of the last domino in the chain, with washing-up liquid detergent. When it falls in the water, it pollutes the surface. The detergent weakens the pull of the surface

tension, where the water adheres to the back of the boat - so that the stronger pull of the unpolluted surface, acting on the front of the boat, can pull the vessel along.

DOMINO ENGINEERING

If you begin by using extra dominoes as props, you can build this arch (Figure 3(a) and 3(b)). The next picture (Figure 3(b)) shows you how...

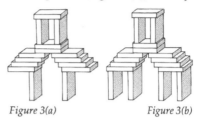

Figure 3(a)   *Figure 3(b)*

If you are thoughtful and patient, you should be able to build these other domino towers (Figures 4 and 5).

A CHALLENGE THAT EXPLOITS STATIC INERTIA

Build the arrangement shown in Figure 6. By

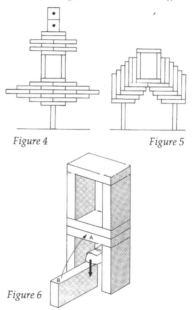

Figure 4   *Figure 5*

Figure 6

poking a finger under the arch, from behind, and pressing on point B, you can flip up the extra domino - to make it hit domino A. If done smartly, domino A should be forced out, without the tower collapsing. (The top part of the tower should just settle down on its legs.)

Alan Ward, 4 Branch Hill Rise, Charlton Kings, Cheltenham GL53 9HW